M000275520

Reading the Fire
– second edition

Compartment Firefighting Series, Volume 2

Shan Raffel AFSM EngTech CFIFireE CF

Benjamin Walker BA(Hons) FIFireE (Godiva Award) MSoA EMT

ISBN: 978-0-6451420-2-0

Contents

About the authors vi

.... viii

Foreword x

Authors' foreword.. xiii

Dedication.. xv

Chapter 1: Revising Fire Dynamics – The foundational principles 1

The development of a fire 1

    The PVT principles 3

Chapter 2: Introducing the 'Size Up' and BE-SAHF 6

Background... 7

    Signs and Symptoms of Flashover 8

    Signs and Symptoms of Backdraft 8

    Signs and Symptoms of Fire Gas Ignition.................... 8

The 4 Key Fire Behaviour Indicators............................ 9

    Contextual Considerations 10

Chapter 3: Smoke Indicators 12

Volume and Location 12

    Smoke location versus fire location........................ 13

Colour... 14

    White Smoke 17

    50 Shades of Grey 18

    Yellowish Brown 21

Buoyancy (descending, drifting, rising)........................ 22

Expansion ("Cauliflowering")................................... 23

Thickness (Optical Density)..................................... 24

Height of Neutral Plane 26

Chapter 4: Air indicators 32

Bi-Directional Flow Path 33

Fuel Controlled – Smooth Flow 34

Ventilation Controlled – Turbulent Flow 35

Uni-Directional Smoke/Air Tracks 35

Uni-Directional Exhaust.. 35

Uni-Directional Inlet 37

Alternating - Pulsing 37

Wind Driven 38

Chapter 5: Heat Indicators 44

Windows and glass 45

Soot stained windows with little or no flame showing...... 45

Cracking or Crazing of Glass. 46

Sudden Heat Build up.. 46

Blistering or discolouration of paintwork...................... 48

Chapter 6: Flame indicators...................................... 53

Volume and Location 53

Colour.. 53

Smoke Auto-igniting 56

Pockets of Flame Forming in the Smoke Layer (Ghosting) 57

Rollover.. 58

Chapter 7: Building Context – Rapid Size up 63

A Complex and Critical Context................................... 63

KISS for rapid fire behaviour size up........................... 64

Flashover 64

Backdraft 66

Fire Gas Ignition 67

Chapter 8: Environmental Context............................... 75

Wind Direction and Velocity 75

Temperature 77

Humidity..77

    Synergistic Effects...77

Chapter 9: Critical considerations for various occupancies-
revisiting the buildings 82

    Introduction 82

    Flashover 82

    Backdraft 83

    Fire Gas Ignition 84

    Wind Driven Fires/Blowtorch Effect 84

Storage, Racking and Compartments within Compartments: . 86

Chapter 10: Be safe, not too late – a small case study...........89

    Step 2 Initial full length unidirectional exhaust flowpath
    (coming out)..93

    Step 3 Uni-directional exhaust continued 95

    Step 4 Ignition of smoke begins just inside the opening....97

    Step 5 Backdraft..99

    Step 6 Approaching peak external involvement........... 101

    The analysis 103

Chapter 10: Your turn at reading fire........................... 106

Chapter 11: Additional types of construction:................... 110

    Considerations for the Initial Incident Commander during Size-
    Up 110

    "Big Box" Type Buildings. 110

    High Rise Buildings (Commercial and Residential)........ 113

    B&B/Boarding Houses (Paying Guest Accommodation) ... 116

    Heritage and Traditional Buildings.......................... 118

    Modern Buildings and Fire Engineering Impacts........... 120

Summary: Building Types:.................................. 122

Chapter 12: Reading the fire – a summary 123

"An average human looks without seeing, listens without hearing, touches without feeling, eats without tasting, moves without physical awareness, inhales without awareness of odour or fragrance, and talks without thinking."

— Leonardo da Vinci

About the authors

Shan Raffel has served as a career firefighter in Brisbane Australia between May 16,1983 to July 16, 2021. His career took a serious change in 1994 after two of his colleagues were killed in the line of duty while conducting fighting operations in a relatively routine fire in a small motorcycle dealership. The coroner's report was unable to identify the cause of two extreme fire events that "caused severe burns, dislocating them from the hose line and rendering them unconscious".

In 1996, two other colleagues were seriously injured after being caught in an extreme fire event while conducting search and rescue operations in a smoke-laden section of a Backpackers Hostel in Rockhampton.

These events motivated him to develop an extensive report that led to an official international study of compartment fire behaviour training (CFBT) in 1997. He studied at the leading training institutions in Sweden and the UK. Over the next three years, he developed the first nationally recognized CFBT program in Australia. Subsequently, he has assisted numerous fire services around the world in the development of their training facilities, instructors, and teaching materials.

In 2009 he was awarded a "Churchill Fellowship" to research "Planning Preparation and Response to Emergencies in Tunnels". This resulted in an intensive ten-week international research tour. This included leading fire services, training centres and tunnel operators in the USA (FDNY), Canada, Germany, Austria, Sweden, Denmark, Norway, and Switzerland. This knowledge was critical in the development of emergency response plans for the 3 largest road tunnels in Australia.

His practical training experience spans 26 countries, and his International Compartment Fire Behaviour Training Instructors program gained international credentialing through the Institution of Fire Engineers recognition process in 2018. The International Tactical Ventilation Instructors course was given IFE recognition in 2022.

www.linkedin.com/in/shanraffel

Since his early retirement from the Fire Service in July 2021, Shan has focused his energy on the development of internationally validated CFBT Instructor development courses. His passion for excellence and safety embraces a broad international network of likeminded professionals.

Benjamin Walker is a globally acclaimed presenter and international compartment firefighting instructor.

He started his career in the Metropolitan Tyne & Wear Fire Brigade, working and commanding some of Europe's busiest fire stations. Following a spell at a small rural fire department, he took a study sabbatical in the United States, obtaining several fire service certifications and studying FEMA's Emergency Management qualifications.

Returning to the UK to train the London Fire Brigade in compartment firefighting, he was recognised by the Institution of Fire Engineers, winning the Godiva Award and subsequent works and contributions led to award of the Fellowship of the Institution aged 40. He has had operational spells at Fire EMS departments in the USA including riding out with Chicago Fire Department, Newport FD and studying Emergency Medicine at Roger Williams University, Providence, Rhode Island.

In demand as a presenter and instructor, he has taught on multiple occasions at the world's largest conference, FDIC in Indianapolis, and remains determined in his pursuit of reducing firefighter line of duty deaths worldwide. He appears as a firefighting subject expert for Sky News, BBC, Russia Today and many other media outlets, writing for The Guardian, FIRE, Fire Risk Management, Fire Engineering and more.

Ben is the Chief and Director of Ignis Global Ltd, a bespoke Fire and Safety Consultancy in the UK, providing training and safety/risk management advice to Fire Services, industry, and private clients. Ignis Global is also an official education provider of ABBE (Awarding Body for the Built Environment) on behalf of Birmingham City University, courses offered include Diploma and Certificate in Fire Risk Assessment, Award in Fire Door Surveying. Fire Safety Auditors (with WFST), Technical Rescue Instructor (with IRRTC) with many more are also available.

Ben and Shan remain at the cutting edge of international operations knowledge and scientific advances. They strive to presenting this essential information in an accessible and clear manner for firefighters wherever they are the world. Stay honest, open, and forthright, without fear.

Foreword

Neil Gibbins, QFSM FIFireE President of the Institution of Fire Engineers 2014/15.

I first met Ben when he attended the IFE AGM to collect his Godiva Award for achieving the highest marks in an IFE examination. I was immediately struck by his energy and passion, which when added to his knowledge (clearly demonstrated by his outstanding exam marks) creates a very powerful persona.

I suspected that we might meet again, and here we are looking at Ben's second book, which he has put together with Shan Raffel. I am delighted to provide a few words for the front of this book, the natural follow up to the first volume, Fire Dynamics for Firefighters. This is so relevant to many operational firefighters today, to help them do their job safely and effectively.

In my formative firefighting years, we would often hear a mantra credited to Massey Shaw, pointing out (I paraphrase poorly) that firefighters need to be able to get inside a burning building, access the fire and use the equipment they have with them to deal with it. In a properly organised fire service, the decision to commit firefighters to the inside of a building on fire is a decision that is only made after the risk has been assessed. To carry out that assessment, the assessor must be equipped with the relevant knowledge to make the defensive/offensive call.

My peers were taught by firefighters who had served in the war years. They might not have been involved in the effects of war, but they worked very long hours and dealt with hundreds of jobs in their careers. They drilled (another old fashioned term) into me what I believed were 'the basics' of firefighting, signs and actions that became normal, second nature. It might sound like firefighting by numbers, indeed it was for training, but it worked, and key signals became embedded so they became almost instinctive. Time moves on, Massey Shaw still holds good, but what has changed?

From the late 50s through to the early 2000s, wholetime promotion required successful completion of a knowledge test. Written exams brought focus to studies; manuals of firemanship were well thumbed and tested. The introduction of the integrated personal development system (IPDS) in 2003, however, saw the end of statutory exams. Thankfully, many UK FRS are utilising some IFE exams, but this isn't the case everywhere. In my first year in the job, my station turned out over 2,000 times. Not a lot, some might say, but today the number for that station is under 500. Spread between four watches, not three.

The point I make is that firefighters in the UK today have less opportunity to deploy their skills, and in many instances have no formal tests of underpinning knowledge or structure for learning. Many, if not most, of the UK officers turning out in charge of fire engines, making decisions to commit firefighters into burning buildings, will, in my opinion, really value Ben and Shan's work. We should celebrate the success of fire prevention and every action that has led to the reduction in fires and deaths but recognise that we are now operating in a new paradigm.

So how does this book (indeed this whole series) help? Ben and Shan translate scientific principles into a form that is easily related to their practical application. They have gathered together the learning about interpreting fires in buildings in a 'modern' context and committed words to paper in 'modern' style. I like reading it. It's not full of jargon, but I think firefighters (especially in the UK and USA) will relate to it and recognise their language. It is broken down into manageable and readable chunks. Ben and Shan tell us what they are going to tell us, do the explanation, and then summarise. This makes the points you need to remember really clear and more likely to be recalled when needed. In the middle of the night, in the rain, with everyone looking to you to make the call. Safely. I thoroughly recommend this book to all students of fire, whether studying to prepare for a firefighting role, an incident command role or indeed any role where the understanding of fire behaviour might lead to better decision making. You will find the words easy to digest and the diagrams complement the words perfectly.

Well done guys! I look forward to the next one. And well done to you for deciding to read it.

Authors' foreword

Thanks for joining us in this second volume of the Compartment Firefighting Series. In this volume, we will be taking a look at how we can effectively use the knowledge acquired in the first book, Fire Dynamics for Firefighters, and apply it to resolve the emergency incidents which we attend in the safest and most effective manner.

Every fire sends out signals that can assist the firefighter in determining the stage of fire development, and most importantly the changes that are likely to occur. This skill is essential to ensure the safest and most efficient firefighting strategy and tactics are employed. Being able to "read a fire" is the mark of a firefighter who is able to make decisions based on knowledge and skill, not guess work or luck.

Shan Raffel, Fire Australia Journal, May 2002, page 11

Just to make the task even more difficult, we are required to make these decisions with incomplete, and often incorrect, initial information. The decisions made will be reviewed by people with unlimited times frames and the benefit of hindsight.

Successful incident size up is like being given two or three hundred pieces of a 1000-piece jigsaw puzzle and being pressured to make an accurate "guess" to the subject of the puzzle in a matter of seconds. The incident commander must make sense of this limited information and develop a plan of action in seconds. As the size-up progresses to a "360", there is potential to find a few more pieces of the puzzle. Additional pieces of the puzzle may be found by vigilant internal teams. The most complete size-up will only be achieved when the "Fire Behaviour Indicators" (FBI) are relayed to the incident commander. Even after a thorough fire investigation, it is not uncommon to find that there are many pieces of the puzzle that cannot be found.

In late 1999, Station Officer Shan Raffel, started to develop a fireground risk assessment model designed to assist in the rapid identification of critical fire behaviour visual patterns. This method is based on the 4 components seen at every fire. Fuel, Air, Heat and Fire. A simple way of thinking of this is to observe the indicators from:

Smoke (think unburnt fuel)

Air Track (flow path)

Heat

Flame

These 4 critical indicators are observed in the context of the building factors and any extremes of environmental factors such as wind velocity, humidity and temperature.

We believe that this method will enhance your ability to make rapid, safe, and effective strategic and tactical decisions. It can assist all your team to identify changes that could indicate potential critical changes in conditions.

If we are not taking actions based on reading the fire, then we are simply acting in a routine and hoping for the best. Hope is not a strategy, so join us now in the challenge to "be safe" by thinking BE SAHF.

Dedication

Krister Giselsson. A Swedish Fire Engineer that had the courage to challenge existing practice, and the intelligence to use science to develop better methods. Truly a man well ahead of his time.

For those that have made the ultimate sacrifice, this has been written to honour your memory, so that others may be safe.

Chris Warburton QFES (Brisbane MFB) 1989

Herbert Fennel and Noel Watson QFES 1994

Jeffery Penfold QFES 2001

Paul Barrow Tyne & Wear Metropolitan Fire Brigade 2010

Billy Vinton Tyne & Wear Metropolitan Fire Brigade 2007

Roy Lewis Tyne & Wear Metropolitan Fire Brigade 2012

"If you stand up and be counted, from time to time you may get yourself knocked down. But remember this: A man flattened by an opponent can get up again. A man flattened by conformity stays down for good."

Thomas J. Watson, Jr

Chapter 1: Revising Fire Dynamics – The foundational principles

"In order to carry on your business properly, it is necessary for those who practice it to understand not only what they have to do, but why they have to do it; and the whole course of my instructions is framed to lead to this end.

No fireman can ever be considered to have attained a real proficiency in his business until he has thoroughly mastered this combination of theory and practice."

Sir Eyre Massey Shaw, KCB "Fire Prevention" 1876

In volume 1 of this series, we covered the stages of development that a fire moves through, and we observed a basic graph detailing this when a fire has sufficient air to sustain its development.

The development of a fire

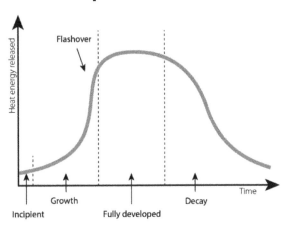

Time temperature curve

We also observed how modern petrochemical based materials, can quickly use up available air to a fire, and how the rate and stage of fire development is limited by the amount of this available air.

In this situation the fire may begin to decay before flashing over. The fire could continue to smolder for a considerable time. As the figure below shows, however, the actions of firefighters 'opening up', ventilating, or even just by their actions in trying to locate a fire such as opening doors while moving through buildings, can all affect the development of a fire to burn and cause the fire to begin developing again.

Unwanted ventilation

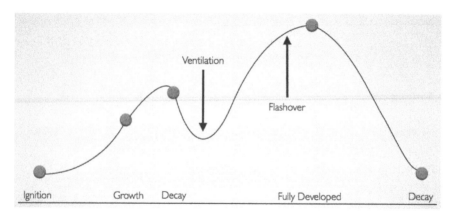

Fire presents extreme dangers and risks and many fire fighters have been caught out by what we term 'rapid fire developments', often caused by this sudden ventilation of a previously ventilation controlled fire. So, let's re-cover what we covered in Volume 1 about these phenomena.

You will recall that we covered a practical model of the three rapid fire developments:

Flashover

Backdraft

Fire Gas Ignition

Feel free to consult your copy of Volume 1 to revise these but let us now look at the signs and symptoms of each.

However, tactically, the approaches and techniques we use will also be governed by the potential for these rapid fire developments and how we can tackle them, maximising fire fighter safety while achieving our goals. Fortunately, we can anticipate (although we can never assume) that some of these developments are more likely in certain situations than others. We will explore these in the following chapters.

The PVT principles

Some final revision involves the gas laws, or, as we defined them in Volume 1, the fire fighter's PVT (pressure/volume/temperature) principles, as related to fire gas/smoke.

The principle states that increasing temperatures will increase the volume of a gas, and that this expansion will continue until a 'container' (such as a closed room's walls, floor, and ceilings) is found. At this point the gas cannot expand much further as it is restricted by the size of the container, and so the volume of fire gas is held constant. If this is the case, and the temperature continues to increase, then the pressure of the gas will also increase – thus we may see pressurised gases trying to escape from any miniscule gaps it can find.

We also know that fire gases will move from areas of higher pressure to those of lower pressure (like letting helium gas out of a balloon as it moves to atmosphere). We can use this knowledge to plan when we create openings which fire gases are going to flow to. Simply opening and closing doors can be a form of 'tactical ventilation' and flowpath management.

To conclude, we need to remember that each flowpath has two aspects – an inlet flowpath for air heading towards a fire, and an outlet flowpath of smoke heading away from a fire. Depending on the opening, these can be 'uni-directional', with separate openings for each flowpath – one in, one out – or they can be 'bi-directional', where one opening is acting as both an inlet opening and exhaust opening. In these cases, we will see fire gas exiting and air entering from the same opening, usually with a clearly defined smoke layer (or neutral plane) between where these different pressures meet.

Bi-directional air track/flowpath

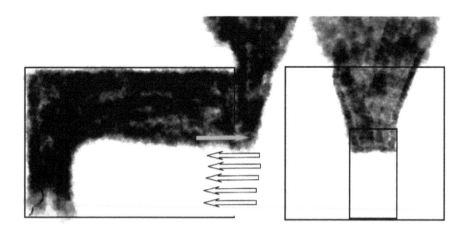

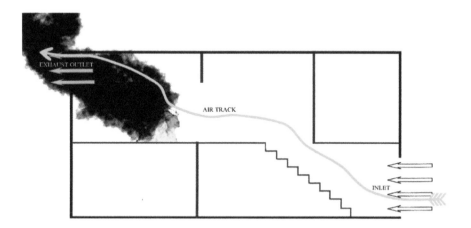

Uni-directional flow

So, wrapping up some brief revision, we should all be back on the page with our terminology, signs and symptoms, gas laws and fire behavior basics. These need to be deeply embedded in our minds so we can anticipate, recognise and act on this information in order to stay safe and achieve our goal – putting fires out!

This chapter covers the following learning outcomes form the Aus-Rescue Pty Ltd International Compartment Fire Behaviour Instructor Level 1, IFE Recognised Training Course.

https://www.ife.org.uk/Training-Development-Directory/142643

Learning Outcome	Description
2.1	Explain fire growth in terms of development phases, burning regimes, flashover, backdraft and fire gas ignition (smoke gas explosions)
2.3	Explain the impact of ventilation openings and the formation of bi-directional and uni-directional flow paths
2.6	Explain the decay phase in terms of fuel or air depletion

The following videos assist with understanding of this chapter- find them at the YouTube channel: tinyurl.com/2xaeb4yu

Video and Hyperlinks:
Pyrolysis and Flashover - https://tinyurl.com/2dysb6de
Fire Profile change to Vent-limited https://tinyurl.com/bdzhjfbz
Laminar Flow vs Turbulent Flow - https://tinyurl.com/rvfknmk4
Bi-Directional Flowpath - https://tinyurl.com/s82y8uu2

Chapter 2: Introducing the 'Size Up' and BE-SAHF

'Size ups' or making scene assessments are as old as the fire service itself. I have no doubt that as ancient Rome burned and Emperor Nero broke out a tune on his violin, some centurion or other, put in charge of this situation, was trying to work out what was happening, what could viably be saved (strategy), what could viably be done with the resources (tactics), and how it could be done (task/operations). I'm sure we've all been in situations where we've had to make an assessment of a similar kind.

The issues that have faced the fire service in making assessments have changed in recent years as we fill our buildings with contents that contain much more energy and have much higher potential heat release rates than 30 years ago, influencing the speed of a fire's spread and its movement through the stages of development.

Furthermore, changes in constructional methods and fixed installations have added more complexity to how a building will or can perform in a fire situation, to both the advantage and disadvantage of firefighters.

Anyone who claims to be able to consistently forecast how a fire will spread by observing smoke is heading for a fall. We work in extremely compressed time frames with incomplete information and rapidly changing conditions. We must realise that some of the fire behaviour indicators are providing us with "soft" or suggestive information. By gathering a number of pieces of soft information we can begin to get a better understanding of the fire location and potential changes.

However, we can still use the knowledge that we acquired in volume 1 to excellent effect by understanding and analysing the dynamics of a fire to answer the following questions with regard to the fire behavior:

- Where is the fire now and where is it heading?

- Who or what is at the greatest risk?

- How can we remove the assets at risk, or can we separate them from the fire?

Most fire departments will have methods of calculating risk, formulating plans and making decisions. However, the information we gather during a 'size up' depends on our knowledge of fire dynamics and our ability to recognise situations and base decisions upon this, known as 'recognition primed decision making'. This knowledge and these skills allow us to gather the most comprehensive information we can and plan effectively with contingencies as required.

Background

The foundations for "Reading the Fire" were laid in Sweden in the late 1970's. Swedish Fire Engineer Krister Giselsson was at the fore front of redefining the accepted knowledge of fire development and extinguishing techniques.

Together with firefighter Mats Rosander, they recognised that changes in building construction and increasing fuel loads from the incorporation of plastics into almost every part of the interior furnishings, was leading to fire development that had never been experienced before.

Using a scientific foundation, they developed practical solutions to educate and equip firefighters with strategies, tactics and tools that could allow them to take back control of the emerging new modern fire ground.

Signs and Symptoms of Flashover

- Lowering of Neutral Plane

- Objects in the room off gassing (Pyrolysing)

- Increased turbulence in the Neutral Plane

- Flow rate of smoke to outside increases

- Flames appearing in the smoke layer

- Rapid temperature rise

Signs and Symptoms of Backdraft

- Fires in enclosed spaces with minimal ventilation

- Signs of an under-ventilated fire

- Oily deposits on windows from the condensation of pyrolysis products on cold surfaces

- Hot doors and windows

- Pulsating smoke gases from small openings

- Whistling sound in openings

Signs and Symptoms of Fire Gas Ignition

- Smoke flows through openings/gaps & accumulates in areas and voids not involved in fire

- Smoke can cool and thin due to mixing with the ambient air

- A lack of heat indicators leads to a false sense of security

The 4 Key Fire Behaviour Indicators

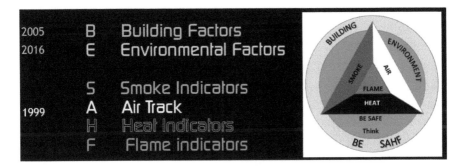

2005	B	Building Factors	
2016	E	Environmental Factors	
	S	Smoke Indicators	
1999	A	Air Track	
	H	Heat Indicators	
	F	Flame indicators	

While I found the indicators listed under the "signs and symptoms" useful for the classroom setting I saw the need for a simple model that could assist in rapidly determining the location of the fire, the stage of fire development and likely changes. In 1999 I was applying a simple mnemonic that focused on the four key fire behaviour indicators: Smoke, Air. Heat and Flame (SAHF "SAFE"). (Raffel, 2002)

Accurate fire development size-up is an essential skill required to develop the safest and most efficient method of attack. A "Tactical Ventilation" plan cannot be safely developed or implemented until a SAHF assessment is carried out.

Tasked teams should also use a SAHF risk assessment in their area of operations. This is particularly critical prior to making any openings. This information should be relayed to the Rapid Intervention Team (RIT) and the IC, so that a more accurate profile of the fire can be developed and maintained.

While the indicators can be read individually, a more complete fire profile can be developed by reading as many of the indicators as possible from the various vantage points inside and outside of the structure.

Contextual Considerations

It is important to realise that there are two critical factors that need to be considered when observing the 4 fire behaviour indicators. The principle factors are the Building and the Environmental variables.

Building
There is a significant amount of information that can assist in the overall strategic size up. This is a large topic, so for simplicity we will focus primarily on building construction from the perspective of the fire behaviour risk assessment. Later in this book we will discuss on how the construction type, and occupancy will influence the indicators likely to be present, and how the fire is likely to progress.

Environment
The most critical consideration is the wind velocity and direction. We will discuss the ground-breaking research on wind driven fires and how other considerations such as extremes of temperature and humidity can influence fire development and the visibility of some of the 4 key fire behaviour indicators.

We will discuss all of these in greater detail later in this book. For now, I want to dive in to how you can identify the 4 fire behaviour indicators and their visual patterns.

This chapter covers the following learning outcomes form the Aus-Rescue Pty Ltd International Compartment Fire Behaviour Instructor Level 1, IFE Recognised Training Course.

https://www.ife.org.uk/Training-Development-Directory/142643

Learning Outcome	Description
5.1	Conduct a Building, Environment – Smoke, Air, Heat, Flame (BE SAHF) assessment to identify fire location, stage of development and likely fire progression
5.2	Develop a strategy from the overall size up, based on a risk versus benefit analysis

The following videos assist with understanding of this chapter- find them at the YouTube channel: tinyurl.com/2xaeb4yu

Video and Hyperlinks:
Conduction, Convection, Radiation- Conduction, Convection- https://tinyurl.com/2y9ttewu
Four Methods of Extinguishing Fire - https://tinyurl.com/4c2pccs4
Fire development curve- https://tinyurl.com/2p95jwzn
Fire Spread convection- https://tinyurl.com/4n5fsaux
Fire Spread radiation -https://tinyurl.com/yckrk8rb
Fire Spread conduction - https://tinyurl.com/ypvzt32d
Fire Dynamics/Fuel Geometry/Flames – https://tinyurl.com/3xvr34ck
Fire Dynamics- changing ventilation profile: https://tinyurl.com/5f88zcxa
Laminar Flow vs Turbulent Flow - https://tinyurl.com/rvfknmk4
Bi-Directional Flowpath - https://tinyurl.com/s82y8uu2

Chapter 3: Smoke Indicators

Smoke is perhaps the most visible indicator that we experience following our arrival upon a scene, and often even on approach to the incident ground. It can give us vital early clues about what's going on in a building.

Smoke indicators can be divided into the following characteristics:

Volume and Location

Colour (white, yellow/brown, grey, black)

Buoyancy (descending, drifting, rising)

Expansion (cauliflowering)

Thickness (light, medium, thick - changes)

Height of Neutral Plane (low, high, descending, rising)

Let's examine these indicators in more detail.

Volume and Location

If we begin our observation on arrival, the first thing that is often apparent is the volume of smoke that presents itself. If a significant quantity of smoke is exiting from the building it can suggest that a fire has spread or that a lot of contents/fuel is involved. A basic principle of smoke is that the more complete the combustion, the less smoke is produced. If the fire is burning efficiently, usually where air is unlimited and readily available, less smoke will be produced than in a ventilation controlled fire where it has a limited air supply.

Smoke location versus fire location

We must remember that smoke may emerge some distance from fire location. Location can be a valuable clue to help us find the firebase, but we must read it in conjunction with the other indicators.

Mushrooming is a term used to describe the horizontal movement of buoyant smoke that has risen to meet an obstruction such as a ceiling. The smoke will move horizontally until it able to find a vertical route or an external opening. If no openings are available, the smoke layer will bank down.

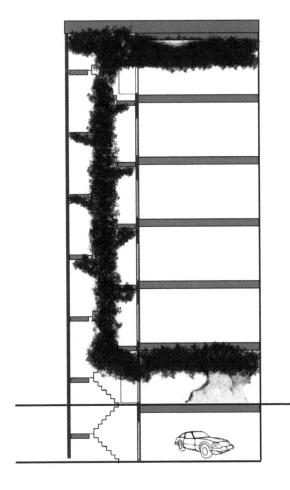

Using our basic scientific knowledge and that precious fire service commodity of common sense, we can assume that higher pressure gas is moving towards areas of lower pressure, and we can use our imaginations to work backwards or 'upstream' against the smoke and make a calculation of where the smoke is originating from.

As previously discussed, fixed installations, shafts, and voids, create routes which can allow smoke to travel and accumulate in some quantity away from the room of the fire's origin. When this happens, the smoke may cool as it travels further from the heat source and pre-mixes with cooler air, and even be diluting into its flammable range. Flame extension could trigger a fire gas ignition.

Colour

Traditional teaching tells us that the colour is determined by the type of the fuel. While this is one of the factors, it overlooks the importance of the influence of the efficiency of the combustion process.

Without a doubt, this is the most poorly understood fire behaviour indicator, and yet it is one of the most important. In order to understand smoke colour, we must know the difference between the following processes:

Pyrolysis: endothermic chemical decomposition from the application of heat energy. (From Greek: pyro "fire" and lysis "separating").

Smouldering Combustion: exothermic surface combustion without flames

Flaming Combustion: gas phase reaction liberating heat and light energy.

PROCESS	PYROLYSIS Decomposition	SMOULDERING Combustion	FLAMING Combustion
DEFINITION	Decomposition of a solid fuel into gasses or vapours. Some will be flammable	Surface combustion without flame, with incandescence and smoke	Rapid reaction that produces light and heat energy
REACTION	Endothermic. External Heat Energy is added, but no energy is released	Exothermic: Once the reaction is started, heat energy is released slowly	Exothermic: Rapid self-sustaining energy release
FUEL	Solids break down to form flammable vapours	Surface combustion oxidation of a porous solid in limited oxygen	Sufficient vapours are formed in the presence of a good air supply
HEAT	100 - 250°C for plastics 220 to 300° for Natural products	Occurs at approximately 600° C or lower	600° C minimum to over 1000° C
OXYGEN	None required!!!!	Small quantities required	Large quantities required

The following 3 screen captures show the effect of heat application to cubes of wood, low density foam, medium density foam and solid plastic. Credits to the "Swedish Fire Nerd". Lars Axleson.

In the very early stages, the process of pyrolysis commences in the low density (yellow) foam. Notice that the colour of the smoke (volatiles) is white.

Later we can see that the low density foam has released all of the volatiles and only the solid char remains.

Eventually the medium density has released all of its volatiles and the solid plastic is just beginning to pyrolyse. Note that the block of wood has hardly decomposed!

- When solid fuels pyrolyse they release white smoke!

- This white smoke has nearly the full potential energy contained in the solid.

- White smoke is VERY high in unburnt fuel content!

- This process does NOT require oxygen

- The plastics found in our modern built environment pyrolyse (decompose) at much lower temperatures than natural products.

- Plastics have to 2.5 to 3.5 more potential energy/kg

White Smoke

So, when we see white smoke there is a good chance this is coming from a space adjacent to, or above the fire compartment. Because pyrolysis occurs at much lower temperatures, the smoke is usually less buoyant. Furthermore, as it moves through corridors and spaces, it will partially mix with air, further reducing the thickness and temperature.

Cold Smoke Can Burn!

This makes cold white smoke on of the most dangerous environments for a firefighter to be in. The low temperature gives a false sense of security. In effect, we have a partially mixed fuel and air layer that can be explosive if an ignition source is introduced!

Flawed Teaching

This is in stark contrast to the teaching of some well-meaning but ill-informed experts who would have us believe that white smoke is "too lean" and has a white colour because it has a high water content. Another mythological explanation is that it started out as black smoke, but when it was pushed through small gaps, all of the carbon got left behind, and as if by magic, it became white. This explanation is also dangerously flawed.

Which brings me to a big caveat regarding white smoke. Once water application has commenced, it becomes difficult to tell the difference between pyrolysis smoke and water laden smoke from water that is hitting the target.

50 Shades of Grey

If the combustion process is very efficient, then most of the soot particles will be burnt in the flame zone. As the air supply becomes restricted, the temperature of the flame will decrease and more of the soot will escape the flame reaction zone.

Thin light grey smoke indicates relatively efficient combustion.

As the combustion process becomes less efficient due to ventilation limited conditions, the amount of soot escaping into the smoke plume will increase along with the volume of smoke.

Thick dark black smoke often indicates fuel rich conditions due to restricted air supply.

In our chapter on flame indicators, we will go into a bit more detail about what different flames colours can tell us. For now, I will summarise the significance of the various shades of grey and the linkage to flame colour.

SMOKE Colour	FLAME Colour	PROCESS
Light Grey - thin	Bright yellow	Fuel controlled
Grey - thicker	Dark yellow to orange	Ventilation limited
Dark Grey - thick	Orange and Red	Ventilation controlled
Black - very thick	Dark red	Severely lacking air

As the smoke colour gets darker, it is highly likely that the percentage of unburnt fuel in the smoke will increase.

In this picture we can see how the simple "4 Compartment Dolls" small scale prop can be used to show the impact of variations in ventilation flow paths. The fire is located in the bottom left compartment. There is an opening across to the bottom right compartment which extends to the top right compartment. (The top left compartment is separated from the other compartments.) After the air supply to the fire compartment (bottom left) has been closed down, flaming combustion ceases, but pyrolysis continues. This produces a white/grey smoke.

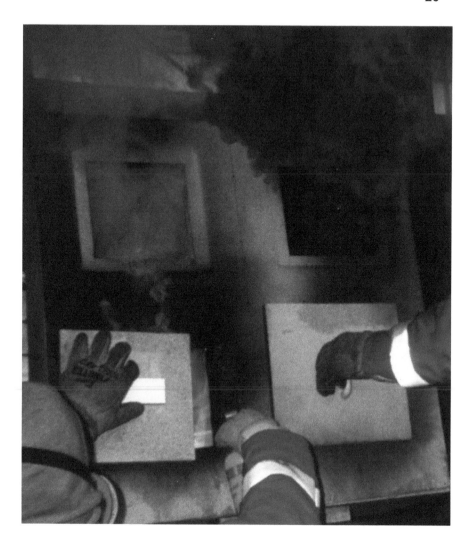

When a slight opening is made in the bottom left fire compartment, a small amount of air is introduced, and this changes the reaction from pyrolysation to a ventilation limited flaming combustion process. The smoke discharging from the exhaust opening at the top right reflects the change in combustion process and the smoke colour changes to black. This is due to the high volume of soot from poor combustion, This will also result in a high percentage of unburnt fuel.

Yellowish Brown

Brown, or yellowish smoke, may be released in the early stages of the pyrolysation of timber products. This is caused when the lignin breaks down and releases tar in aerosol form.

In the modern built environment, most of the contents are petrochemical based and do not often produce a brown smoke. One area that contains mostly timber combustibles, is the roof space of dwellings. Timber roofing materials have evolved from solid timber elements that are nailed together, to light weight timber held together with "gang nails". This increases the relative strength to weight ratio, but often compromises the fire resistance rating.

Buoyancy (descending, drifting, rising)

Smoke at high temperatures will be lower in density (buoyant) and will rise and expand aggressively and rapidly.

Smoke that is lower in temperature will rise slowly or even drift downwards.

Expansion ("Cauliflowering")

I must admit that "Cauliflowering" is not actually a scientific term. It has been adopted from a fantastic YouTube video of a UK instructor describing the indicators leading up to a backdraught (AKA backdraft) in a large-scale demonstration.

I include this "term" because it is a very practical way of giving an image of this visual pattern indicator.

Smoke seen expanding as it leaves the room, indicates that it has been slightly compressed. Compartment fires can produce pressures in the vicinity of 15 pascals to over 20 pascals. The expansion on leaving the exhaust opening may indicate that his opening is close to the fire seat.

This clip shows expanding smoke with colours ranging from light to dark grey.

Thickness (Optical Density)

The thickness can be a good indicator of the efficiency of the combustion process. In the early fuel controlled stage, the rate of smoke production is less due to relatively good air supply.

As the fire transitions from fuel controlled, to ventilation controlled, the combustion process become less efficient. This results in an increase in the percentage of unburnt fuel, soot, and smoke volume. The neutral plane lowers and becomes so optically dense, that it becomes impossible to

Smoke thickness is also a function of duration. So, it is possible for a relatively small fire to produce a large volume of thick smoke, if it has been burning for a long time in ventilation limited conditions.

Height of Neutral Plane

These photos show the lowering of the neutral plane as the fire progresses towards flashover. Photo courtesy Tim Watkins

As the fire develops the neutral plane will lower and the thickness of the smoke gases will increase.
Therefore:

- A high neutral plane could indicate that the fire is in the early stages of development.

- A very low neutral plane could indicate very rich backdraft like conditions.

- A sudden rise could indicate that ventilation has occurred.

- Gradual lowering could indicate a build-up in fire gases and an approaching flashover.

- Sudden lowering could indicate a sudden intensification of the fire.

Summary

In general, smoke colour gives us an indicator of how efficient the combustion process is:

- White – pyrolysis products.

- Light Grey – good to fair combustion so a reasonable air supply.

- Medium Grey – ventilation limited combustion.

- Dark Grey – poor ventilation supply.

   - Black – very ventilation limited but still burning or at least smouldering.

> For firefighters, what do smoke conditions tell us?
>
> Smoke location is not a reliable indicator of fire location but can present clues. We must remain aware that 'travel' may have taken place.

Dark smoke is generally an indicator of limited ventilation to fires and can signpost those gases have moved above their flammable range into a 'too rich' concentration of fuel to air.

White/lighter colour smoke can be fuel-rich pyrolysis products created by non-flaming combustion in a fire compartment, or the heat from the fire compartment causing pyrolysis in adjacent areas – classic examples being floors above basements. This indicates a strong potential for rapid fire development (fire gas ignition).

Neutral plane positioning can indicate the burning profile and stage of fire development plus the potential for rapid fire developments.

Remote auto ignitions of smoke indicate high temperatures and a rich concentration of fire gases, so oxygen/air must be a critical consideration.

Changes in air supply (air admitted or restricted) can affect smoke colour.

Fire spread can be indicated by changes in smoke colour.

Smoke can present differently in different locations even in small buildings.

Fixed installations operating, like sprinklers, can alter visible smoke/fire gases and conceal the true nature of the fire.

Which indicators should we be looking for with smoke?

Location/volume.

Colour.

Buoyancy.

Thickness (optical density).

Neutral plane levels.

AIDE MEMOIRE: SMOKE

INDICATOR	FLASHOVER	BACKDRAFT	FIRE GAS IGNITION
VOLUME AND LOCATION Varies with the room size, geometry, openings, duration of fire, and air supply.	May be unreliable as a single indicator. Must be read with the other indicators that are present in the fire compartment itself.	Large volumes of smoke will be concentrated in the fire compartment. Other parts of building may have a variety of smoke conditions. Unreliable unless read with other indicators.	Smoke can emerge and accumulate some distance from the source. This can give a false indicator of the location of the fire compartment.
COLOUR - Varies with fuel: Type (natural or synthetic) Form (gas, liquid, solid, shavings, dust) FIRE PROFILE: -fuel controlled, or -ventilation controlled	Requires a good air supply. Will vary from grey to dark grey. As flashover approached it may become ventilation controlled which will darken the colour towards black.	White smoke may indicate the contents are undergoing pyrolysis with limited flaming and smouldering. Yellow/brown can indicate decomposition of wood. Black will generally indicate at least active smouldering of very limited flaming combustion.	Smoke that has travelled some distance from the fire compartment may appear lighter in colour due to partial mixing with cooler air as it moves through the structure.
BUOYANCY hot smoke rapidly rises	Buoyancy increases as the compartment approaches flashover. Smoke emerging some distance from the room	Buoyant, expanding smoke indicates higher internal temperature and pressure - possibility of AIT. Less buoyant (lazy smoke) indicates	Generally, low buoyancy due to premixing with the cooler ambient air.

	of origin will be less buoyant.	lower temperature but is still dangerous!	
THICKNESS	Becomes thicker as flashover approaches.	Usually, thick. When at a high temperature it will expand and roll at a high velocity.	Can often appear to be thinner (to some extent) due to pre mixing with cooler, fresh air.
HEIGHT OF NEUTRAL PLANE smoke layer air interface	Pre-flashover height is high and rapidly lowers as flashover becomes imminent.	Low or at floor level in the compartment of origin. 'Bouncing' neutral plane can indicate this too.	Usually not well defined due to premixing with cool air.

This chapter covers the following learning outcomes form the Aus-Rescue Pty Ltd International Compartment Fire Behaviour Instructor Level 1, IFE Recognised Training Course.

https://www.ife.org.uk/Training-Development-Directory/142643

Learning Outcome	Description
1.3	Explain complete combustion, incomplete combustion and passive agents
1.4	Describe flammable limits and the impact of variations in temperature and pressure.
2.1	Explain fire growth in terms of development phases, burning regimes, flashover, backdraft and fire gas ignition (smoke gas explosions)
2.3	Explain the impact of ventilation openings and the

	formation of bi-directional and uni-directional flow paths
2.5	Explain fire spread through multi compartment structures
5.1	Conduct a Building, Environment – Smoke, Air, Heat, Flame (BE SAHF) assessment to identify fire location, stage of development and likely fire progression

The following videos assist with understanding of this chapter- find them at the YouTube channel: tinyurl.com/2xaeb4yu

Video and Hyperlinks:
Pyrolysis of Fuel- https://tinyurl.com/5n7n2jdu
Smoke is Fuel - https://tinyurl.com/4d9c5z4v
Flammable Range - https://tinyurl.com/2evtuu8t
Flammable Range 2 https://tinyurl.com/bderf775
Flowpath Management - https://tinyurl.com/4tmkyfph
Small Scale Demonstrations - https://tinyurl.com/bddrj7sc
https://www.youtube.com/watch?v=WG-YCpAGgQQ&list=PLJlr-eg3qfqGS9YgQTfYNkYdZYlcIylm8&index=7
https://www.youtube.com/watch?v=FgMinHC44DU&list=PLJlr-eg3qfqGS9YgQTfYNkYdZYlcIylm8&index=6

Chapter 4: Air indicators

We have already touched upon the influence of air supply, or lack of it, upon fires and our ability to read them. Put simply, the presence of air is going to have a huge effect on the fire condition, its profile (fuel or ventilation controlled) and our tactical decision making.

To revise, again, flowpaths can be 'inlet flowpaths' taking air supply to the fire, or 'outlet flowpaths' taking fire gases (smoke) products of combustion away from the fire. They can be 'bi-directional', with the same opening acting as both inlet and outlet, or 'uni-directional', where air comes in through one opening, and smoke/fire gas exits through another. There can also be elements of both, primarily one or the other. If we examine this, taking a simple approach, let's look at what is going in (air) and what is coming out (smoke/fire gas).

Air is one of the four fire behaviour indicators that may be observed at an opening or within a structure. Air being invisible is typically characterized through the movement of the contrasting smoke, its velocity, turbulence and height within the boundaries of a compartment or at an opening. When combined with the smoke indicators it is known as the smoke/air track.

It is vitally important that firefighters understand the likely pathways of air movement due to planned or unplanned ventilation in all parts of the structure.

To allow for a rapid initial assessment, I have divided the Air Indicacators into 3 basic types, each with 2 sub types

Bi-directional

- Fuel Controlled Pattern

- Ventilation Controlled Pattern

Uni-directional

- Inlet Pattern

- Exhaust Pattern

Alternating uni-directional

- Pulsing Pattern

- Wind Driven Pattern

Bi-Directional Flow Path

When an opening is created in the room on fire, the heated gases will flow out of the top of the opening and cool air will be drawn in through the bottom of the opening.

The Bi-directional smoke/air track can be divided into 2 sub-groups. These are "fuel controlled" and "ventilation controlled".

Fuel Controlled – Smooth Flow

Slow and smooth flow could indicate that the fire is in the early stages and most likely still fuel controlled. This statement is reliable if the smoke we are reading is discharging directly from the compartment on fire. However, if the fire compartment is some distance from where we are witnessing the release, the smoke velocity may reduce as it flows through the non-involved compartments. If the distance is large it I is possible that even a ventilation controlled fire could show a smooth flow path after extended travel.

Ventilation Controlled – Turbulent Flow

As the fire grows larger the demand for air increases as does the volume of super-heated smoke. Eventually this leads to a situation where the discharged smoke blocks off the opening and restricts the ingress of air towards the fire based. This results in turbulence at the neutral plane in the opening.

Uni-Directional Smoke/Air Tracks

Uni-Directional Exhaust

Smoke or flames seen leaving an opening from the very bottom to top of the opening in an indicator of a total exhaust outlet. For the smoke or flames to be flowing directly out of an opening, there must be inlets that are at least equal in cross sectional area or wind being driven into the inlet/inlets.

Uni-Directional Inlet

Look for open doors and windows that could be feeding the fire. As previously mentioned, each Uni-Directional exhaust opening will require inlets of at least equal cross sectional area (unless the inlets are wind driven). It may be possible to reduce the rate of fire development if these inlets can be closed or restricted until hose lines can be put in place.

Alternating - Pulsing

Smoke seen pulsing out of small openings can indicate a ventilation controlled (or limited) fire. This indicates that there are variations in pressure due to limited oxygen supply. As the oxygen level decreases so does the combustion process. This in turn decreases the temperature and consequently the volume of the gases decrease. This causes air to be drawn in, the fire starts to increase, the temperature of the gases increases, and the expanding smoke is pushed out through the gaps under pressure until the air is consumed and the cycle starts again. In some cases, this could develop into a situation where the sudden opening of the compartment could lead to backdraft. Whistling noises may indicate that air is being pushed in and out of the compartment through small gaps or openings due to pressure variations. This indicates a ventilation controlled fire. It should be remembered that it might be difficult to notice this with background noise.

Wind Driven

Under normal wind conditions, a room with only one opening will display a Bi-Directional air track. This will be either fuel controlled (smooth flow) or ventilation controlled (turbulent). Strong winds being blown into a closed fire compartment can lead to a high pressure zone in the compartment. In the wind driven scenario, the opening will aggressively alternate from a total inlet to a total exhaust outlet.

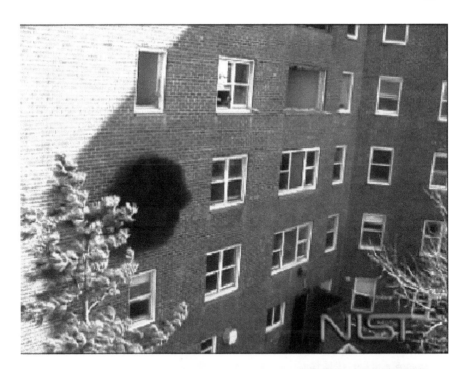

Strong wind overcomes the pressure of the fire and injects a charge of air into the room

This intensifies the fire to the point where the pressure is greater in the room than the pressure created by the wind

This belches out the smoke/flame – blocks off air intake – creating a cycle

---

For firefighters: what do air indicators tell us?

The presence of air inlets can indicate whether a fire has a 'fuel controlled' or 'ventilation controlled profile'.

No visible air inlets may suggest a 'ventilation controlled' situation, but should be assessed alongside the available air that is already within a building, open plan layouts etc.

Multiple air inlets can cause turbulence of fire gases and rapid fire developments.

If we can identify a single opening acting as an air inlet and smoke outlet (bi-directional flowpath) we may be able to use this as an exhaust vent if we create another inlet opening and attack via that route.

If a uni-directional flowpath is already created (one air inlet, one smoke/fire gas outlet) we can use this to our advantage and increase the speed of attack.

The 'kill zone' or outlet side of the flowpath can be identified and crews briefed, and tactics made accordingly.

'Sucking' of air at air inlets can be signs of vent controlled profiles, creating negative pressures and backdraft potential.

'Belching' can suggest a 'sealed', wind-driven fire and tactics can be altered to avoid 'Blowtorch' effects, as covered in Volume 1.

AIDE MEMOIRE AIR INDICATORS:

INDICATOR	FLASHOVER	BACKDRAFT	FIRE GAS IGNITION
FLOW PATH (Smoke & Air Tracks) Bi-Directional (Bi-D) Uni-Directional (Uni-D) Wind Driven (WD)	Smooth Bi-D air tracks indicate fuel controlled conditions. Turbulence at the neutral plane indicates a ventilation controlled regime. Neutral plane will be descending.	Less apparent due to limited number or size of openings. Look for openings alternating between Uni-D inlet to a UniD outlet on initial opening. Smoke is often roiling and expanding.	The further the smoke moves from the compartment or origin, the more likely the interface will be smooth and appear benign!
PULSATIONS	Not seen in the fuel controlled phase but may present to some extent in the ventilation controlled phase.	Often present. Rapid pulsations usually mean higher temperatures.	Highly Unlikely
WHISTLING SOUNDS	Unlikely	Pulsing air forced in and out of small gaps may make this sound.	Highly Unlikely

This chapter covers the following learning outcomes form the Aus-Rescue Pty Ltd International Compartment Fire Behaviour Instructor Level 1, IFE Recognised Training Course.

https://www.ife.org.uk/Training-Development-Directory/142643

Learning Outcome	Description
1.6	Explain how heat energy can be transferred via conduction, convection and radiation
2.3	Explain the impact of ventilation openings and the formation of bi-directional and uni-directional flow paths
2.4	Explain the impact of a Wind Driven fire on the heat release rate
3.3	Explain ventilations tactics and techniques that can be synergistically combined with various nozzle techniques
3.4	Explain compartment entry procedures/techniques and the concept of the kill zone, buffer zone and safe zone
5.1	Conduct a Building, Environment – Smoke, Air, Heat, Flame (BE SAHF) assessment to identify fire location, stage of development and likely fire progression
5.2	Develop a strategy from the overall size up, based on a risk versus benefit analysis
5.5	Explain how ventilation tactics/techniques can be combined with suppression tactic/techniques to increase the overall safety and efficiency

The following videos assist with understanding of this chapter- find them at the YouTube channel: tinyurl.com/2xaeb4yu

Video and Hyperlinks:
Wind Driven Fires (Jerry Tracy) https://tinyurl.com/2p8srbj5
Wind Driven Fires (Symptoms) https://tinyurl.com/j7c27ws8
Compartment Entry- https://tinyurl.com/2s3dcvk5
Control Air- Control Fire https://tinyurl.com/5699nsh4 Backdraft Symptoms- https://tinyurl.com/yckz8afa
Backdraft- Domestic Property - https://tinyurl.com/y6ra4ca7
Reading Fire- https://tinyurl.com/bde3pkyk
Flowpath Management - https://tinyurl.com/4tmkyfph
Small Scale Demonstrations - https://tinyurl.com/bddrj7sc

'It is most dangerous for any persons who happen to be in other rooms of the house, particularly those above and at the back, into which, after a front window has been cut through (broken), it is probable, if not almost certain, that the fire will penetrate before the firemen can reach them...'

Sir Eyre Massey Shaw 1868

Chapter 5: Heat Indicators

Heat indicators become very important when there is limited smoke showing. Building construction is a major factor in which heat indicators are likely to be evident. Heavy, well insulated construction is less likely to show heat indicators in the early stages.

Heat signatures can often be tricky to spot, and it is here where recent(ish) technology has become our friend. Thermal image cameras can greatly assist us during an initial size to identify clues that heat presents. And, lucky you, we will be covering the tactical use of thermal image cameras in our 'toolbox' section of this book.

This image taken during UL/NIST research gives an excellent comparison of the power of TI during size up. There are very few heat indicators visible with the human eye, but the TIC reveals temperature in excess of 600°F (300°C)

All items have a 'thermal capacity'. It will absorb heat and act as a 'passive' agent until it reaches this thermal limit, or capacity, and begins to contribute that heat back into a compartment. Items such as bricks have high thermal capacities, and this heat can be identified by thermal imaging. We can use this to our advantage during 'size up'.

Windows and glass

Soot stained windows with little or no flame showing

The blackening indicates rich conditions ventilation controlled conditions indicating potential for backdraft. This is often accompanied with oily deposits on the inside of the window.

Windows can also become distorted by heat, and surrounding fixtures, particularly those consisting of polymers or plastics, can melt or warp, giving another indicator of particularly intense heat presence.

Cracking or Crazing of Glass.

Glass is susceptible to 'thermal shock'. Rapid heat build-up can result in cracking of glass. Furthermore, sudden reductions in heat can causes cracks or failures so we should exercise caution when applying water to avoid making unwanted openings and creating more flowpaths.

Glasswork can show more of a 'crazing' (fine cracks) effect where the heat buildup has been slower and steadier. This is often accompanied with blackening and oily deposits and indicates high temperature fuel rich conditions.

Credit Shan Raffel

Sudden Heat Build up

This is frequently quoted as an indicator that flashover or backdraft is impending. If you are waiting for this indicator, it is likely that you have missed the other fire behaviour indicators and are in a lot of trouble! This is a late indicator as it often occurs AFTER some form of fire gas combustion (rollover) has commenced in the ceiling area. This may be difficult for the firefighter to see and will not give adequate warning time. By the time the firefighter in modern PPE senses the temperature increase, they will be in a very dangerous situation. Never rely on sensing sudden heat build-up.

One of the best tools for monitoring heat buildup is the Thermal Imaging Camera (TIC). To accurately interpreting the image, it is essential that the user has a sound understanding of the essentials of thermal imaging technology as well a working knowledge of the strengths and limitations of their particular camera. For those of you that are looking to increase your operational efficiency in the use of your TIC, I recommend that a solid starting point for your education can be found at Insight Training LLC. www.insighttrainingllc.com. Andrew Starnes and his highly experienced team, provide a wealth of knowledge that is built on the latest scientific research, frontline experience, international collaboration, and their own practical live fire experiments.

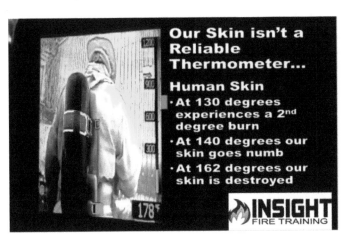

Credit Andy Starnes, Insight Fire Training

Actions can be taken as precursors to this stage such as training 'temperature checks' by water application into smoke layers, which we shall cover in our 'toolbox' section

Blistering or discolouration of paintwork

Blistered paintwork can also be an indicator of heat build-up. Painted surfaces on lightly insulated doors and walls (with lower thermal capacity) can begin to bubble and blister. It is perhaps as likely that a change in colour of paintwork that is exposed to heat will be more visible to the naked eye.

Heat damaged paintwork. Credit Shan Raffel

Heated surfaces can present a number of patterns which we can add assist in locating the seat of the fire. As with all indicators we shouldn't be too hasty into making assumptions. A surface may not show signs of heat, but could contain fire gases within their flammable range, waiting for an ignition source (fire gas ignition). As we know, as all available oxygen is used up within ventilation controlled fires, burning can cease, heat release reduces and temperatures can fall, so situations with the potential for backdraft may not always show signs of heat.

Initially, feeling the door surface and door handle may help to detect developing heat conditions.

Sweeping a water spray across a lightweight door or surface can also be used to test for surface heat. If the door is over 100C, the film of water on the upper section will rapidly evaporate. In some cases, it is possible to get an indication of the height of the neutral plane by observing the line at which the evaporation ceases. Heavily insulated doors will not show this effect.

For firefighters, what do the heat indicators mean to us?

Firstly, and very importantly, we should be looking for areas of heat using a thermal image camera wherever possible.

Severe heat in one area can indicate proximity to fire, however as the heat travels through cooler air, indicators may be less obvious, but the risk of fire gas ignition, with fire gases diluted into their flammable range even though they're at lower temperatures, is increased.

Severe blackening of windows indicates where hot fire gases are 'too rich' to burn, meaning the concentration and flammability limits are crucial and air tactics should be relevant and appropriate.

Heat signatures will give an indication of fire location, particularly if the fire profile is fuel controlled.

Bricks and other materials with higher thermal capacities will absorb and retain heat but may take longer to warm up than lightweight materials that soon reach their capacity and become 'active' and 'emitters'.

AIDE MEMOIRE: HEAT INDICATORS

INDICATOR	FLASHOVER	BACKDRAFT	FIRE GAS IGNITION
PAINTWORK BLISTERED OR DISCOLOURED Heat indicators are less obvious structures with heavy insulation	Often visible with lightly constructed uninsulated doors	Discoloration may give indication of heat layering	A lack of heat indicators could be deceptive. Smoke cools and pre-mixes as it travels
WINDOWS Soot Stained Condensate Cracking or crazing	Windows may crack if the heat build-up is sudden (or if jets are applied to super-heated glass)	Darkening indicates rich conditions. Cracking = rapid temperature changes. Crazing indicates gradual temperature increase rate	Darkening may be present as the smoke accumulates. Cracking is less likely in the early stages
HOT SURFACES (May be absent in structures with heavy insulation)	Often present in lightweight construction	Surfaces may be hot, & the temperature will decrease as the available oxygen is consumed	May not be hot, particularly in the early stages
SUDDEN INCREASE IN INTERIOR TEMPERATURE	A very late indicator and therefore of no use in giving early warning! Gas cool and/or remove the smoke.	A very late indicator and therefore of no use in giving early warning. Anti-ventilate and/or cool smoke before ventilation & have charged hose line in place!	None until the fire gas ignition. which can be very sudden and even explosive. Explosive power depends on the amount of fuel and how well it has pre-mixed with the available air.

This chapter covers the following learning outcomes form the Aus-Rescue Pty Ltd International Compartment Fire Behaviour Instructor Level 1, IFE Recognised Training Course.

https://www.ife.org.uk/Training-Development-Directory/142643

Learning Outcome	Description
1.3	Explain complete combustion, incomplete combustion and passive agents
1.4	Describe flammable limits and the impact of variations in temperature and pressure.
2.1	Explain fire growth in terms of development phases, burning regimes, flashover, backdraft and fire gas ignition (smoke gas explosions)
2.2	Explain the factors which affect the development and spread of a compartment fire including geometry, linings, and fuel package location
2.3	Explain the impact of ventilation openings and the formation of bi-directional and uni-directional flow paths
5.1	Conduct a Building, Environment – Smoke, Air, Heat, Flame (BE SAHF) assessment to identify fire location, stage of development and likely fire progression
5.2	Develop a strategy from the overall size up, based on a risk versus benefit analysis
5.3	Develop and implement an incident action plan that will preserve life/property/environment that is in compliance with the "safe person concept"
5.4	Explain the need for continual dynamic risk assessment, taking into account changes in E SAHF

	indicators and ongoing hard/soft information
5.5	Explain how ventilation tactics/techniques can be combined with suppression tactic/techniques to increase the overall safety and efficiency

The following videos assist with understanding of this chapter- find them at the YouTube channel: tinyurl.com/2xaeb4yu

Video and Hyperlinks:
Backdrafts- https://tinyurl.com/p7836we9
Fire Gas Ignition - https://tinyurl.com/4nxpddn7
Flashover- https://tinyurl.com/bdh5hddx
Reading Fire- https://tinyurl.com/bde3pkyk
Water Application- https://tinyurl.com/yn2h928s
Flowpath Management - https://tinyurl.com/4tmkyfph
Small Scale Demonstrations - https://tinyurl.com/bddrj7sc

Chapter 6: Flame indicators

There is a tendency for firefighters to focus on any visible flame on arrival at a scene. There is nothing inherently wrong with this if it does not lead to a loss of situational awareness. However, unless the structure is totally involved, it is still critical to read all the fire behaviour indicators to get a complete picture on the current stage of fire development, and the likely path of fire extension.

Volume and Location

Incidents in which flames are externally visible on arrival obviously make it easier to determine the seat of fire and likely direction of spread. It is important to look for signs of multiple seats of fire and to realise that visible flame may have spread some distance from the original source.

Colour

Traditional teaching tells us that the colour of the flame can give an indication of the particular product that is burning. While this is true in situations where a single product is burning, it is important to realise that the same product can burn with different coloured flames depending on the combustion process. For example, LPG that is premixed with air will produce a blue coloured flame (due to the presence of CO_2). If the fuel and air are mixed by the process of diffusion, then the flame will be yellow due the presence of carbon particles from a less efficient combustion process. LPG burning in an oxygen deficient, or fuel rich environment can produce a red flame.

The following photos show the different flame colour that LPG can produce when there are variations in the fuel concentration and the amount of pre-mixing.

A simple example of a solid fuel is the combustion of

particleboard in a compartment. When the air supply is good it will burn with a yellow flame. If the oxygen concentration is reduced the flame becomes a reddish orange colour.

In a compartment fire, **yellow flames** generally indicate a reasonable air supply. As the combustion process becomes less efficient (less oxygen) the flames with start to turn **orange** and then **red.**

There is a crude rule of thumb that says a flame burns brightest when it is hottest, but we must remember that the availability of air, the type of fuel and pyrolysis products generated can all influence this.

It is perhaps regarding the influence of air that we can begin to analyse flame colours.

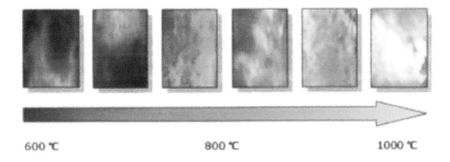

600 ℃   800 ℃   1000 ℃

Credit (Lambert & Baaj)

The shape or form of the flame can also give an indication of the type of combustion occurring. The reddish orange flames that result from the rich combustion are often turbulent with a short wave form. The ignition of accumulated pyrolysis products produces a very light yellow flame, sometimes almost clear. Amazingly in this case, the wave form is larger, and the flames will move downwards seeking the higher oxygen concentration. As with all of the indicators in the SAHF assessment, it is important to observe the initial flame colour and then note any changes.

Smoke Auto-igniting

Smoke auto-igniting outside an opening indicates that the internal conditions are above the auto-ignition temperature (AIT) and too rich to support full flaming combustion inside the compartment. When the super-heated rich smoke leaves the compartment, it is able to dilute down form a mixture within the flammable range. If it does not cool down to below the AIT during this process, the fuel content can ignite spontaneously. When these conditions are encountered it is critical to realise that increasing the air available to the room will result in a sudden and possibly violent increase in fire intensity (backdraft). Under these conditions doors or windows should be closed until hose lines are in place. Careful application of water into the space before ventilation can reduce the temperature of the smoke to below the AIT and reduce the likelihood of sudden

intensification.

Pockets of Flame Forming in the Smoke Layer (Ghosting)

If internal crews begin to see pockets of flame forming above the neutral plane this is an indicator that the unburnt fuel in the smoke layer is approaching the AIT. Cool the smoke and consider withdrawal.

Rollover

Once the accumulated unburnt fuel begins to ignite, it is common for the flames to roll across the ceiling resulting in a rapid increase in radiative heat. This could lead to flashover or fire gas ignition. The application of water in the ceiling space may stop or delay the progression of the rollover. If it is not possible to place water on the burning surfaces to slow the heat release rate, the crew should retreat to a point of safety.

For firefighters, what are the flames saying to me?

Flame colour can indicate the fire profile (fuel or ventilation controlled).

A yellow flame can indicate good combustion and hotter temperatures. As combustion becomes less efficient the colour will change from orange to red.

Flames in the smoke/fire gas layer are an indication of impending flashover.

Lack of visible flames in conjunction with previous BE-SAHF factor, can indicate potential for rapid fire developments.

AIDE MEMOIRE FLAME INDICATORS:

INDICATOR	FLASHOVER	BACKDRAFT	FIRE GAS IGNITION
LOCATION & VOLUME	Caution! Visible flame may have spread some distance from the initial seat of fire. Look for the seat and/or multiple points of origin	Little or no visible flame, but conditions can vary widely in different parts of the structure. Super-heated fuel rich smoke may auto ignite after leaving the compartment of origin	No flame in adjacent areas prior to FGI. After ignition it is likely to be progress very rapidly (or explosively). Cool the gases and/or remove accumulated smoke!
AUTO IGNITION: (REMOTE IGNITION OUTSIDE OF AN OPENING)	Indicates that the internal conditions are extremely hot (above AIT) and too rich to support full flaming combustion inside the compartment	May be seen occurring as pulsed smoke is pushed out. May also occur after opening the compartment. Can trigger backdraft	Less likely to trigger a FGI than an ember or flame extension.
FLAMES FORMING IN THE SMOKE	Isolated flames traveling in the hot gas layer. May not be visible without a TIC. Indicates	Could occur after ventilation if water is not introduced.	Not Likely

LAYER	impending rollover and flashover.		
ROLL OVER. Ignited fire gases roll across the ceiling leading to a massive increase in heat flux	This is a late indicator of flashover and it not always visible to the naked eye. Use your TIC and prevent it with Gas Cooling.	May occur after ventilation it the backdraft is not triggered early by auto-ignition or embers.	Roll over in the compartment of origin could progress into the smoke accumulated in adjacent areas and trigger an FGI.
COLOUR (Can be influenced by many variables)	Yellow colour at the base of the fire often indicates good air flow. Reddish orange flames may indicate less air is available or the conditions are fuel rich.	Red or orange may indicate fuel rich conditions. Pockets of blue flames are said to be from the auto ignition of carbon monoxide.	No flame may be present in the space prior to ignition.

This chapter covers the following learning outcomes form the Aus-Rescue Pty Ltd International Compartment Fire Behaviour Instructor Level 1, IFE Recognised Training Course.

https://www.ife.org.uk/Training-Development-Directory/142643

Learning Outcome	Description
1.3	Explain complete combustion, incomplete combustion and passive agents
1.4	Describe flammable limits and the impact of variations in temperature and pressure.

1.5	Explain the chemistry of combustion in solids, liquids, gases, dusts and vapour phases
2.1	Explain fire growth in terms of development phases, burning regimes, flashover, backdraft and fire gas ignition (smoke gas explosions)
2.2	Explain the factors which affect the development and spread of a compartment fire including geometry, linings, and fuel package location
2.3	Explain the impact of ventilation openings and the formation of bi-directional and uni-directional flow paths
2.5	Explain fire spread through multi compartment structures
5.1	Conduct a Building, Environment – Smoke, Air, Heat, Flame (BE SAHF) assessment to identify fire location, stage of development and likely fire progression
5.2	Develop a strategy from the overall size up, based on a risk versus benefit analysis
5.3	Develop and implement an incident action plan that will preserve life/property/environment that is in compliance with the "safe person concept"
5.4	Explain the need for continual dynamic risk assessment, taking into account changes in E SAHF indicators and ongoing hard/soft information
5.5	Explain how ventilation tactics/techniques can be combined with suppression tactic/techniques to increase the overall safety and efficiency

The following videos assist with understanding of this chapter- find them at the YouTube channel: tinyurl.com/2xaeb4yu

Video and Hyperlinks:
Complete Combustion v Incomplete Combustion https://tinyurl.com/ycy46zdm
More combustion profiles https://tinyurl.com/4b7y4tmn
Candle Experiments https://tinyurl.com/2sh24xw6
Water Application- https://tinyurl.com/yn2h928s
Flowpath Management - https://tinyurl.com/4tmkyfph
Small Scale Demonstrations - https://tinyurl.com/bddrj7sc
Backdrafts- https://tinyurl.com/p7836we9
Fire Gas Ignition - https://tinyurl.com/4nxpddn7
Flashover- https://tinyurl.com/bdh5hddx
Reading Fire- https://tinyurl.com/bde3pkyk

"A fireman, to be successful, must enter buildings; he must get in below, above, on every side, from opposite houses, over brick walls, over side walls, through panels of doors, through windows, through loopholes, through skylights, through holes cut by himself in the gates, the walls, the roof; he must know how to reach the attic from the basement by ladders places on half burned stairs, and the basement from the attic by rope made fast on a chimney. His whole success depends on his getting in and remaining there and he must always carry his appliances with him, as without them he is of no use."

Sir Eyre Massey Shaw 1868

Chapter 7: Building Context – Rapid Size up

A Complex and Critical Context

The building construction type can give us an indicator to what indicators may be present and how the fire is likely to develop. Indicators that are prominent in one type of construction may not be present in another.

A good example of this can be found in the heat indicators. Buildings that have HVAC or the modern trend in the construction of "Passive House" design that utilises a range of features that significantly improves energy efficiency. The principal components are highly energy efficient insulation, ventilation, and construction design. This combination makes the heat indicators more difficult to recognise.

The critical factors are complex. Some of the key factors include:

- Occupancy
- Construction materials and type
- Fuel load, form, and type

This is a subject that could fill an entire book and it is well beyond the scope of this volume. As a firefighter that is seeking excellence, it is essential that you educate yourself on the types of buildings that are commonly found in your patch.

I would like to provide you with an iron clad decision making model that will consistently allow you to assess all of these factors in a few seconds. Unfortunately, this is impossible. There is not enough time and there is not enough information.

So should we just respond like robots and use our default

tactics, tools, and techniques. Or stick our thumb in our mouth and assume a foetal position?

I am certain we can do better than either of these options. Building and construction used to be a fundamental aspect of firefighter training and it is important to know how buildings are put together, and how they can fall apart. Elements of constructions such as beams, columns, floors and walls can all react differently in fires and affect the stage of development as well as the potential for rapid fire developments.

As a professional firefighter, it is your responsibility to be an expert in the building construction that is common in your patch. Armed with this knowledge you have a solid foundation to make rapid building assessments

KISS for rapid fire behaviour size up

My initial arrival, and 360 Building assessment, focusses on the fire location and the most likely fire spread scenario. At the simplest level, we should try to make a rapid assessment on the most likely mechanism of fire progression. Flashover, Backdraft or Fire Gas Ignition

Flashover

We know that in order for flashover to occur, there must be a sufficient air supply into the compartment. There must be large compartments or openings that will allow for air supply. It is common for buildings in warmer climates to have open plan layout and large single panes windows. Even if the windows are closed there is some airflow, and it is common for these large sheets of glass to crack when exposed to rapid heat change. Many "modern" window frames are made from plastics or aluminium. These materials will lose their integrity at relatively low temperatures.

These examples of a typical "Queenslander" timber home that was popular in the late 19th century early 20th century, in the tropical and sub-tropical regions of north eastern Australia.

Timber framing, internal and external cladding, floors, ceiling and even the stumps. Large verandas, large windows, high ceilings and other features designed to make the most of the breeze in the long hot summers.

Do you have buildings in your patch that have similar features that allow ample air for rapid development and fire spread?

Backdraft

Building construction that limits the supply of air during the growth stage may result in a very gradual decay before flashover is achieved.

Examples of construction features include:

- Small compartments with well-sealed doors and windows

- Double or triple glazed windows

- Heavy insulation

Some of these features may be found in buildings with HVAC systems. The "Passive House" construction features high insulation factors and minimal air leakage.

In these construction types the fuel packages will cease free flaming combustion as the oxygen levels drop below 16%. Smouldering may continue in low oxygen concentrations for extended periods. Remember, that solid fuels will continue to pyrolyse even in the absence of oxygen, provided that the temperatures are somewhere between 120 to 350C.

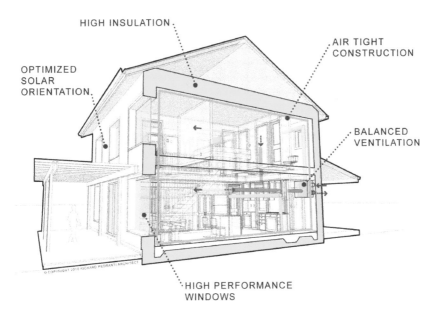

Energy efficiency Passive Home concepts. Credit Richard Pedranti Architect

Fire Gas Ignition

"Fire gases lying in wait"
Krister Giselsson

Most structures we are called to deal with have multiple compartments on each level. A fire in one room my allow smoke to flow to uninvolved parts of the structure. We know that this results in some mixing with air, which, may thin the smoke and reduce the temperature. In addition to the normal corridors in building we may and service connections between different compartments and levels. If these that not been professionally fire stopped, the smoke may accumulate in unexpected places.

So, what are the key construction features we need to look for?

- Occupancies that have opening such as ducting or conveyor systems

- Buildings that have undergone renovations

- Buildings with common roof spaces

- Suspended ceilings

In 2000 Shan attended a fire in a restaurant that had started in a small storeroom. The structure was originally used as a small picture theatre. A suspended ceiling was added to what was originally the seating area. An addition had been added to the rear of the structure which was being used as the main kitchen and pantry.

The fire had taken a hold of the below ground level on arrival. The fire was being attacked from the rear when the author tasked a BA team to make entry with a hoseline from the front street level to check for fire extension on that level. Approximately 5 minutes later the author joined the BA team.

They had entered in through the dining area and found little or no smoke. Buffer zoning of this area had been conducted and the team had begun to check the preparation area. There was very little smoke or heat in this area and no sign of fire on first observation. About 3 metres down the central passageway, the SAHF assessment revealed a small room the showed signs of heat.

The door was opened using the standard entry technique. The sight that met the team immediately started alarm bells ringing. Due to poor construction, the fire from below had passed largely undetected through the wall between the weatherboards and the gypsum plaster internal wall lining. Having burnt through a section of the internal lining into this room, the flame and smoke had travelled along the ceiling straight into an open access hatch to the ceiling space. This discovery alerted the team to the fact that the large ceiling void had been filling with smoke and heat for several minutes. At this time ventilation of the roof had not been completed due to problems caused by the location of power lines, roof design and limited access for aerial appliances.

3D Water-fog was gently introduced in an effort to cool the gases without entraining air. It was becoming apparent that

there had been a number of additions and alterations to the

structure and that these had created a number of interconnecting voids. The author called for external ventilation on the front roof gable, and the team then began a very cautious retreat while continuing to pulse and paint. The team had not proceeded far when a smoke explosion occurred in the ceiling void.

The force blew down the suspended ceiling and resulted in rapid fire spread through the roof void. At the time of the explosion, the team was in the shelter of a doorway and were able to

proceed to safety. The buffer-zone that had been created on entry had helped to prevent immediate involvement of the dining area. Subsequent water application actually stopped the progress of fire into that area and allowed the team to exit the building.

"As a fire officer, you should always remember that you will be judged on the outcome of your decisions, not by what was happening at the time"
Person: Battalion Chief John Kincaid, Abilene Fire Dept., TX.

Aide memoire: Building

BUILDING FACTOR	FLASHOVER	BACKDRAFT	FIRE GAS IGNITION
TYPE OF CONSTRUCTION Construction materials, thermal properties & geometry have an enormous influence on development & structural stability. USE AND OCCUPANCY Indication of the likely life risk, fuel type/load STAGE OF DEVELOPMENT FIRE PROFILE STRUCTURAL INTEGRITY	Flashover will occur in most buildings if sufficient air is available. Lightweight single pane windows & doors may fail allowing sufficient air for flashover to occur. Large open plan with limited compartmentation will allow rapid spread. Heavy brick or cement rendered walls will absorb a lot of energy which could delay flashover.	Backdraft is more likely in energy efficient buildings with good insulation and sealed openings such as double/triple glazed windows. Developing fires may consume the available oxygen before it can transition to fully developed. Heat indicators may be less obvious due to the superior insulation associated with this type of construction.	Voids, ducts, shafts, open plan, false or suspended ceilings etc. allow smoke to be transported & accumulate in areas adjacent to, or some distance from the compartment of origin. Poor or damaged smoke/fire stopping may be found in original or modified buildings. Unburnt fuel is partially mixed with fresh air and can accumulate to flammable concentrations. Conducted heat may pyrolyse combustible elements in adjacent spaces

This chapter covers the following learning outcomes form the Aus-Rescue Pty Ltd International Compartment Fire Behaviour Instructor Level 1, IFE Recognised Training Course.

https://www.ife.org.uk/Training-Development-Directory/142643

Learning Outcome	Description
2.1	Explain fire growth in terms of development phases, burning regimes, flashover, backdraft and fire gas ignition (smoke gas explosions)
2.2	Explain the factors which affect the development and spread of a compartment fire including geometry, linings, and fuel package location
2.3	Explain the impact of ventilation openings and the formation of bi-directional and uni-directional flow paths
2.5	Explain fire spread through multi compartment structures
2.6	Explain the decay phase in terms of fuel or air depletion
5.1	Conduct a Building, Environment - Smoke, Air, Heat, Flame (BE SAHF) assessment to identify fire location, stage of development and likely fire progression
5.3	Develop and implement an incident action plan that will preserve life/property/environment that is in compliance with the "safe person concept"
5.4	Explain the need for continual dynamic risk assessment, taking into account changes in E SAHF indicators and ongoing hard/soft information

The following videos assist with understanding of this chapter-
find them at the YouTube channel: tinyurl.com/2xaeb4yu

Video and Hyperlinks:
Construction Effects on Fire Behaviour-
Backdrafts- https://tinyurl.com/p7836we9
Fire Gas Ignition - https://tinyurl.com/4nxpddn7
Flashover- https://tinyurl.com/bdh5hddx
Flowpath Management - https://tinyurl.com/4tmkyfph
Small Scale Demonstrations - https://tinyurl.com/bddrj7sc
Reading Fire- https://tinyurl.com/bde3pkyk

Chapter 8: Environmental Context

In the context of the key environmental factors that need to be considered during our rapid BE SAHF risk assessment, wind direction and velocity is the most critical. In this chapter we will also briefly discuss the impact of temperature and humidity.

Wind Direction and Velocity

UL and NIST have conducted extensive research in the impact of wind on fire development. The results have provided us with an appreciation of just how powerful the effect can be. Ben will discussing this in greater detail later in this book, but for now we will focus on the rapid sizeup considerations.

Firefighters can start the observation of wind velocity and direction well before arrival. When conducting the 360 size-up it is critical to observe the conditions on all 4 sides as the terrain, vegetation and surrounding built environment can change the direction and velocity of the wind. The factors that cause this effect are complex and beyond the scope of this book. For a detailed explanation of we highly recommend the latest edition of Fire Ventilation by Stefan Svensson.

Urban or built up environments can affect the wind in two ways: buildings can be shielded from wind by other buildings, and deflections from taller buildings can add frictional drag, causing turbulence, eddies, and areas of low pressure on the 'leeward', or protected, side of the building. Wind striking on the windward side can result in increased pressure zones and deflection of the wind direction. The combination can cause significant changes in wind direction and velocity.

Being aware of these low pressure areas is vital given what we know from Volume 1 about fire gases moving from areas of high pressure to areas of lower pressure. Furthermore, wind direction and speed are critical when planning ventilation operations – this is explored further in Fighting Fire, Volume 3 of the Compartment Firefighting Series.

Buildings that are fairly evenly spaced can also create wind channels or tunnels where the wind moves at faster speeds, a version of the venturi effect in a way (uninterrupted by buildings and turbulence). For example, the grid system of buildings in Chicago creates these wind tunnels where the wind from Lake Michigan can pick up speed and is amplified.

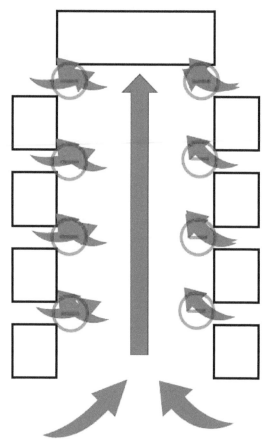

These situations can lead firefighters to encounter high speed winds and we must account for them in our assessment and the effect that it can have on fires.

Temperature

Following many discussions and consultations, we found that in order for external temperatures and conditions to start having a visible effect on conditions, we would have to be operating in extremely cold conditions, so this may only really apply to those of you reading in Alaska, Canada, Scandinavia and other colder climates. For the rest of us, unless there is a huge spectrum and difference in temperatures between the warmest you encounter and the lowest, then the impact on the appearance of BE-SAHF indicators is unlikely to be noticeable.

Edinburgh University's Fire Dynamics faculty were kind enough to succinctly state: 'Cold temperatures won't have a major effect on smoke plume appearance.'

(@Edinburgh Fire twitter account)

Humidity

Moisture in the air or atmosphere will also have a minor effect upon fire development, acting as a passive agent and absorbing heat until it reaches its thermal capacity, higher levels of humidity in the atmosphere will reduce the ability of moisture in the fuel to readily evaporate to air. As a very crude rule of thumb, the higher the humidity, the harder it is for fuel to cure.

Synergistic Effects

The combination of very low humidity, and high temperatures, can have a powerful effect. When this is combined with high wind velocity, the impact can be extreme! Extended high temperature, wind and low humidity with dry out (cure) vegetation and timber.

Wind Driven Fires | NIST

FIRE FIGHTING TACTICS UNDER WIND DRIVEN CONDITIONS:
LABORATORY EXPERIMENTS (highrisefirefighting.co.uk)

What considerations are relevant to environmental conditions?

Geographical location – consider what environment you are operating in (urban/rural) and how this will impact on response times for reinforcements, availability of water supplies and proximity to other risks? All of these influence our tactics and size up.

Wind impact – velocity and direction – remember that conditions may lend themselves to the potential for wind driven fires.

Check for signs and symptoms of wind driven fires being exhibited ('belching' fire gases).

Analyse whether the landscape could amplify the effects of the wind, such as in Chicago's 'wind channels'.

Note which is the upwind side (leeward) and therefore safest to attack from if possible.

Consider the requirement for specialised equipment (e.g., floor below nozzles).

Very low temperatures means that air has a higher oxygen content, so extra caution should be exercised when managing flowpaths or admitting air after creating openings.

Very low humidity alone will not have a significant impact on development or visible FBI, but it can increase the impact of a wind driven fire.

INDICATOR	ALL RAPID FIRE DEVELOPMENTS (Flashover, Backdraft, FGI)
Wind Velocity & Wind Direction	This factor can play a major role in the rate and direction of fire spread. The velocity will magnify the effects of low humidity and extremes of temperature. The natural lay of the land or the built environment around the structure can cause variations in direction and velocity of the wind on the various sides of the involved structure. The 'wind driven fire' scenario is an example of how external wind conditions can have an extreme effect on fire behaviour. It is critical to consider wind direction and velocity before making any openings.
Low Humidity	Prolonged low humidity conditions can dry vegetation to a 'cured' state. This increases the potential ignitability of surrounding vegetation and accumulated leaf litter in gutters and around the outside of the structure. A similar effect can influence the rate of ignitibility and spread of structure components and even contents to lesser extent. This effect is exacerbated by high temperatures and high wind velocity.
Extreme Heat	Fighting operations can become extremely arduous in these conditions and frequent relieves combined with core cooling may be essential to rehabilitate crews. Extra vigilance is required to prevent heat stress and dehydration.
Extreme	Rapid cooling of discharged smoke may decrease

Cold	the buoyancy. When combined with 'low atmospheric pressure' an inversion layer can form that will prevent the smoke from rising.
	The construction commonly found in areas likely to experience extreme cold are less likely to show heat indicators that could be considered 'common, normal or reliable' in tropical/sub-tropical climates.
	Firefighter welfare in these conditions requires knowledge and SOP's based on experience. This is business as usual for departments that commonly experience extreme cold.
	The highest risk exists when these conditions occur as part of a 'rare extreme' event in areas that are not experienced in dealing with extreme cold

'Rescue is always our highest priority at a structure fire, but it should not be the first thing we do unless, of course, we are determined in getting ourselves injured or killed! Therefore I always maintain that the best way to rescue people from a burning building is to put out the fire'. Chief Shapher challenged those who disagreed with him to read the NIOSH reports to see how firefighters get injured or killed whilst making rescue attempts.'

F. Shapher - Chief of St. Charles MO

This chapter covers the following learning outcomes form the Aus-Rescue Pty Ltd International Compartment Fire Behaviour Instructor Level 1, IFE Recognised Training Course.

https://www.ife.org.uk/Training-Development-Directory/142643

Learning Outcome	Description
1.4	Describe flammable limits and the impact of variations in temperature and pressure.
2.1	Explain fire growth in terms of development phases, burning regimes, flashover, backdraft and fire gas ignition (smoke gas explosions)
2.3	Explain the impact of ventilation openings and the formation of bi-directional and uni-directional flow paths
2.5	Explain fire spread through multi compartment structures
5.1	Conduct a Building, Environment – Smoke, Air, Heat, Flame (BE SAHF) assessment to identify fire location, stage of development and likely fire progression
5.3	Develop and implement an incident action plan that will preserve life/property/environment that is in compliance with the "safe person concept"
5.5	Explain how ventilation tactics/techniques can be combined with suppression tactic/techniques to increase the overall safety and efficiency

Chapter 9: Critical considerations for various occupancies- revisiting the buildings

Introduction

During the First Edition of Reading Fire, we perhaps got a little too bogged down in the explanations of building, construction and environmental effects.

Whilst it is a fundamental skill of each and every Firefighter to understand buildings, how they are built, how they burn and the dangers of collapse, we perhaps went slightly off course in Edition 1 rather than just focus on the essentials that can be evaluated in a micro-second as part of an effective "size up" using the SAHF factors. In response to that critique, we will assess the buildings in response to the potential for rapid fire developments that they present.

This section also shows the importance of local knowledge and knowing one's "ground" from the Firehouse.

Flashover

Simply, the presence and availability of an air source is the requirement for a compartment to proceed to Flashover. In terms of these buildings, larger volume sized compartments, open-plan offices all have a plentiful supply of air, either within, or accessible from the outside.

"A" factors- such as multiple inlets, broken windows in semi-derelict or run-down properties are also signs that there is potential for a fire to develop to flashover. Similarly, lightweight or badly insulated buildings can often fail and create additional inlets and outlets that may assist the move towards Flashover.

Contents and fuel should be considered in all building types. Remember that the heat flux and speed of development is related to the energy transfer. Modern contents are high energy products. With a good supply of air, this is a perfect recipe to move a fire to flashover.

Backdraft

Whilst a building needs air to develop to flashover, a lack of air can help create backdraft conditions. Buildings that lend themselves to lack of air are normally well-insulated with smaller internal compartments. In the past, we have often thought about the colder-weather nations such as the Scandinavian countries having great building insulation and lighter construction types being in warmer nations, but this can be misleading. There are many extremely warm countries now that have excellent insulation properties in the buildings in order to keep occupants cool.

Today's glass products are excellent insulators, and although we can check for other signs and symptoms, thermal signatures, a modern building that has copious glazing can present as high a risk of backdraft as a traditional smaller compartment building.

"S" factors, as well as "A" should be considered in tandem with looking at the building type. A lack of potential inlets, the known use of highly insulating building materials, should all be considered here.

Heavy, high-energy content loads (fuel) coupled with small, air restricted compartments will create a condition where heat causes the fuel to release its pyrolysis products and unburned fire gases, but without fire, the energy isn't removed from these in the form of heat and light- so these rich, energy full gases are ready, when opening up, to dilute (with air) back down into the flammable range and ignite, potentially even detonate.

Fire Gas Ignition

Voids, Gaps and Hollows. Natural voids creating in the design of the building such as attic spaces or in roofs, plus those artificially created in compartments by use of "drop-down" or "false ceilings" in Office and Commercial styles of buildings can lead to travel away from the fire of origin or accumulation of unburned fire gases/pyrolysis products and the potential for fire gas ignitions.

Lightweight construction with cavities, timber framed cavities and buildings known to have issues with compartmentation integrity all have FGI potential.

The conversion of former industrial or commercial buildings, such as cinemas, bonded warehouses and factories may contain certain areas that lend themselves to FGIs such as shafts, former conveyors or chutes moving between compartments.

Whilst modern building codes and regulations should prevent this and a thorough inspection and report compiled prior to being "signed off", this is not always the case. At time of writing in 2021, Ignis Global Ltd and our partner fire-stopping company Saracen Compliance surveyed and rectified a building in Central Birmingham, England that had in excess of 400 "penetrations"- or gaps leading to voids, a lot of potential for travel of buoyant fire gases to other compartments, awaiting an ignition source.

Wind Driven Fires/Blowtorch Effect

Wind Driven Fires remain a dangerous potential for all those engaged in high-rise firefighting operations. Explained in "Fire Dynamics", these are a matter of understanding the external signs and symptoms such as "belching" and managing the interior flowpath effectively, not getting caught on the wrong side of a uni-directional flowpath in the "kill zone" or "fatal funnel".

Rapid Fire Development:	Building Clues:
Flashover:	Consistent and good supply of air, either from the exterior (broken windows etc, or the volumetric size of the compartment and air contained within. High energy contents can lead to excellent and fast combustion with a good supply of air.
Backdraft:	Restriction of air from external and internal sources. Generally smaller compartment sizes within the building. Excellent insulating properties may also be a clue in extremely hot or cold weather nations. High energy contents and restricted air will lead to the release of fuelled pyrolysis products that haven't lost their energy as lack of complete combustion has not caused these to be released as heat and light.
Fire Gas Ignition:	Voids, hollows and gaps. Converted buildings where the interior layout has been altered and new walls, false ceilings and cosmetic alterations have been made in a renovation phase. Remember that these fire gases will alter in their flammable range as they

	move away from their source or state of origin and could be considerably more dangerous when a significant distance away from the original fire.
Blowtorch/Wind Driven:	High Rise Buildings, with known corridors/lobbies and approach routes.

Storage, Racking and Compartments within Compartments:

Certain modern stores such as Costco, to use one example, have high level racking in a warehouse format, organised into channels. To a lesser extent, supermarkets also have this format. If the shelving or racks are full, we can create "tubes", "tunnels" and compartments within the compartment.

Fires can be at different stages of development throughout such large premises known as "Travelling fires", some of which are structurally designed to collapse into its own footprint in the event of fire. Incident Commanders should be acutely aware of the potential for this and familiarise with such buildings and businesses on their station ground.

'It has been said that there is as much difference between a man who has not trained and cultivated his intellect and one who has, as between a dead man and a living, and the same contrast may be made between those who have not studied fire brigade work and those who have.'

Sir Eyre Massey Shaw 1868

This chapter covers the following learning outcomes form the Aus-Rescue Pty Ltd International Compartment Fire Behaviour Instructor Level 1, IFE Recognised Training Course.

https://www.ife.org.uk/Training-Development-Directory/142643

Learning Outcome	Description
1.4	Describe flammable limits and the impact of variations in temperature and pressure.
2.1	Explain fire growth in terms of development phases, burning regimes, flashover, backdraft and fire gas ignition (smoke gas explosions)
2.3	Explain the impact of ventilation openings and the formation of bi-directional and uni-directional flow paths
2.5	Explain fire spread through multi compartment structures
5.1	Conduct a Building, Environment – Smoke, Air, Heat, Flame (BE SAHF) assessment to identify fire location, stage of development and likely fire progression
5.3	Develop and implement an incident action plan that will preserve life/property/environment that is in compliance with the "safe person concept"
5.5	Explain how ventilation tactics/techniques can be combined with suppression tactic/techniques to increase the overall safety and efficiency

The following videos assist with understanding of this chapter- find them at the YouTube channel: tinyurl.com/2xaeb4yu

Video and Hyperlinks:
Construction Effects on Fire Behaviour- https://tinyurl.com/2p85ytc4
Backdrafts- https://tinyurl.com/p7836we9
Fire Gas Ignition - https://tinyurl.com/4nxpddn7
Flashover- https://tinyurl.com/bdh5hddx
Flowpath Management - https://tinyurl.com/4tmkyfph
Small Scale Demonstrations - https://tinyurl.com/bddrj7sc
Reading Fire- https://tinyurl.com/bde3pkyk

Chapter 10: Be safe, not too late – a small case study

We're now going to take a brief look at an incident in a BE-SAHF context. Let's review what signs we can identify in the images below for each of the areas we've looked at, and consider, not criticize, whether we would make different decisions here.

The following situation demonstrates what can happen if we don't pay proper attention to the BE-SAHF principles or the indicators of rapid fire developments such as backdraft.

This chapter consists of a series of images taken in quick succession as a pair of firefighters attempt to gain access to a burning building. Each image is followed by a table highlighting the important points that can be observed and that should be taken into account in similar situations.

So, let's get started!

Step 1: Uncontrolled Vent Creation (Can't be closed/shut)

'The men of the fire brigade were taught to prevent, as much as possible, the access of air to the burning materials. What the open door of the ash-pit is to the furnace of a steam-boiler the open street door is to the house on fire. In both cases the door gives vital air to the flames.'

...the door should be kept shut while the water is being brought, and the air excluded as much as possible, as the fire burns exactly in proportion to the quantity of air which it receives.

James Braidwood Superintendent, London Fire Brigade

Fire Prevention and Fire Extinction 1866

BUILDING	We have a large area of single pane glass. Expect rapid fire growth if windows fail.
Flashover	
Backdraft	If this is a shop there is a potential for high fire loading and for oxygen to be used up quickly, heightening the risk of backdraft.
FGI	
	There will potentially be voids allowing fire gases to accumulate, which raises the risk of FGI.
	Consider whether there is a basement and if the fire could be located there. It is possible that, due to lack of ventilation, it is creating lots of pyrolysis products or

	heating the floor above causing FGI potential on the ground floor.
ENVIRONMENT Air direction and velocity	No noticeable signs of wind impact.

SMOKE
 	
  Volume/Location
 	
  Colour
 	
  Buoyancy
 	
  Height of neutral plane
	There is little smoke visible at this time from this perspective.
	
	The dark colour indicates ventilation limited combustion. So, ventilation (adding air) will most likely intensify the fire.

AIR
 	
  Bi-directional
 	
  Uni-directional
 	
Pulsations/Alternating flow path
	The opening created with the axe appears to be creating a uni-directional inlet.

HEAT
 	
  Blistering or discoloration of paintwork
 	
  Cracking or crazing of windows
 	
  Condensate, tar, or soot on windows
 	
  Hot surfaces
 	
  Sudden increase in
	Blackened windows, heat distortion/cracks – could be a sign of backdraft conditions.

temperature	
FLAME	Nonvisible from this angle currently.
Location and Volume	
Auto-ignition	
Pocketing in smoke layer	
Rollover	
Colour (red to orange to yellow)	

Step 2 Initial full length unidirectional exhaust flowpath (coming out)

BUILDING	No change or additional information.
Flashover	
Backdraft	
FGI	
ENVIRONMENT	No change or additional information.
Air direction and velocity	
SMOKE	Smoke emerging low and as a uni-

Volume/location	directional exhaust.
Colour	Smoke is a dark colour indicating ventilation limited combustion.
Buoyancy	
Height of Neutral Plane	Buoyant and expanding low neutral plane.
AIR	Change to uni-directional exhaust is a classic backdraft warning sign.
Bi-directional	
Uni-directional	
Pulsations/Alternating flow path	
HEAT	Blackened windows, heat distortion/cracks- (Backdraft?)
Blistering or discoloration of paintwork	
Cracking or crazing of windows	
Condensate, tar, or soot on windows	
Hot surfaces	
Sudden increase in temperature	
FLAME	Nonvisible from this angle at this time.
Location and volume	
Auto-ignition	
Pocketing in smoke layer	

Rollover Colour (red to orange to yellow)	

Step 3 Uni-directional exhaust continued

BUILDING Flashover Backdraft FGI	No change or additional information.
ENVIRONMENT Air direction and velocity	No change or additional information.

SMOKE Volume/location Colour Buoyancy Height of neutral plane	Volume and velocity increasing rapidly. Dark colour of smoke suggests rich combustion. Buoyant and expanding fire gases. Very low neutral plane.
AIR Bi-directional Uni-directional Pulsations/Alternating flow path	Uni-directional exhaust is increasing in velocity and volume – backdraft warning sign.
HEAT Blistering or discoloration of paintwork Cracking or crazing of windows Condensate, tar, or soot on windows Hot surfaces Sudden increase in temperature	Blackening rapidly increasing.
FLAME Location and Volume	Nonvisible from this angle at this time.

Auto-ignition Pocketing in smoke layer Rollover Colour (red to orange to yellow)	

Step 4 Ignition of smoke begins just inside the opening.

BUILDING Flashover	No change or additional information.

Backdraft FGI	
ENVIRONMENT Air direction and velocity	No change or additional information.
SMOKE Volume location Colour Buoyancy Height of Neutral Plane	Volume and velocity increasing rapidly. Buoyant and expanding fire gases. Very low neutral plane.
AIR Bi-directional Uni-directional Pulsations/Alternating flow path	Uni-directional exhaust is increasing in velocity and volume – backdraft has commenced!~~warning sign.~~
HEAT Blistering or discoloration of paintwork Cracking or crazing of windows Condensate, tar, or soot on windows Hot surfaces Sudden increase in	Blackening rapidly increasing.

temperature	

FLAME
 	
   Location and Volume
 	
   Auto-ignition
 	
   Pocketing in smoke layer
 	
   Rollover
 	
   Colour (red to orange to yellow)
	The transformation of smoke to flame (red to orange) has commenced just inside the doorway.

Step 5 Backdraft

BUILDING
 	
  Flashover
 	
  Backdraft
 	
  FGI
	No change or additional information.

ENVIRONMENT Air direction and velocity	No change or additional information.

SMOKE
 	
  Volume/location
 	
  Colour
 	
  Buoyancy
 	
  Height of Neutral Plane
	Volume and velocity still increasing.
	
	Lighter colour indicates improved combustion efficiency.
	
	Buoyant and expanding (hot).
	
	Very low neutral plane.

AIR
 	
  Bi-directional
 	
  Uni-directional
 	
  Pulsations/Alternating flow path
	Uni-directional exhaust of smoke transitions to flame.
	
	Backdraft progresses outside of the opening.

HEAT
 	
  Blistering or discoloration of paintwork
 	
  Cracking or crazing of windows
 	
  Condensate, tar, or soot on windows
	Transition from hot smoke to flame.
	
	Temperature rapidly increasing.

Hot surfaces	
Sudden increase in temperature	
FLAME	Yellow to orange flame commencing at the base of opening, where there is the best air supply.
Location and Volume	
Auto-ignition	
Pocketing in smoke layer	Triggered by auto-ignition as it comes into contact with the air outside the building.
Rollover	
Colour (red to orange to yellow)	

Step 6 Approaching peak external involvement

BUILDING Flashover Backdraft FGI	No change or additional information.
ENVIRONMENT Air direction and velocity	No change or additional information.
SMOKE Volume/location Colour Buoyancy Height of neutral plane	Volume and velocity increasing rapidly. Buoyant and expanding. Very low neutral plane.
AIR Bi-directional Uni-directional Pulsations/alternating flow path	A total uni-directional exhaust approaching peak intensity Well-developed backdraft.
HEAT Blistering or discoloration of paintwork Cracking or crazing of windows Condensate, tar, or soot on windows	Transition from hot smoke to flame increasing.

Hot surfaces Sudden increase in temperature	

FLAME
 	
   Location and Volume
 	
   Auto-ignition
 	
   Pocketing in smoke layer
 	
   Rollover
 	
   Colour (red to orange to yellow)
	Yellow to orange flame progressing.
	
	White flame indicates very high temperatures in lower part of the opening.

The analysis

So how do the actions in the photographs relate to the stages of fire development? In short, if we revisit Volume 1 and have a look at Chief Ed Hartins' diagram of the classic ventilation induced fire development curve leading to backdraft.

We can see that the crucial factor is the point of increased ventilation. In this case, caused by the breaking of the windows in step 1.

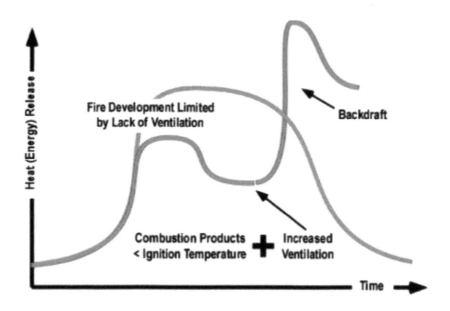

(CFBT US- Ed Hartin)

Classic Ventilation Induced Fire Development Curve leading to flashover

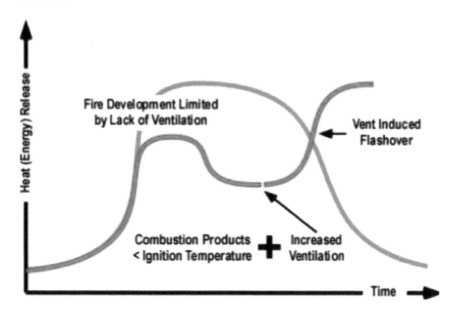

CFBT US- ED Hartin

For firefighters, what do these photos and graphics this show us?

It is absolutely critical that the BE-SAHF assessment takes place as an overall and related assessment. The timing of making the assessment is crucial and must take place before any tactical operations take place.

The photos used above were taken from a time frame of less than 15 seconds, so we can see how rapidly things can change.

Some questions to consider:

Would you use the same tactics as the crew?

If not, what tactics would you use instead?

Don't worry to too much if you can't answer the second question yet. Volume 3 in this series will give you a number of options, tactics and techniques to put into play.

Chapter 10: Your turn at reading fire

So, having covered the basics of making a BE-SAHF assessment, it's over to you to undertake an assessment of each of the following images using the knowledge you've picked up. There's nobody watching, so feel free to refer back to previous chapters and your notes here!

We would suggest printing out one copy of the BE-SAHF assessment form at the end of the chapter for each of the images and complete it to the best of your ability.

IMAGE 1:

Image 2 (www.sheboyganpress.com)

IMAGE 3: (Robert Busbee)

BE-SAHF assessment sheet

INDICATOR	ON ARRIVAL
Building Flashover Backdraft FGI	
ENVIRONMENT Air direction and velocity Extremes of temperature or humidity	
SMOKE Volume/Location Colour Buoyancy Height of Neutral Plane	
AIR Bi-directional Uni-directional Pulsations/Alternating flow path	
HEAT Blistering or discoloration of paintwork Cracking or crazing of windows	

Condensate, tar, or soot on windows Hot surfaces Sudden increase in temperature	
FLAME Location and Volume Auto-ignition Pocketing in smoke layer Rollover Colour (red to orange to yellow)	

Chapter 11: Additional types of construction:

Considerations for the Initial Incident Commander during Size-Up

The Incident Commander must form an immediate picture of the situation upon arrival, and whilst there is always danger in over-reliance on "what should happen", any excessive delays can result in "paralysis by analysis" and the time frame for an effective size-up, brief and initial actions lost.

To reiterate earlier messages, we must "know our patch", the buildings within it and what they contain (fuel loading). The following categories offer a generic and immediate appraisal of some common factors that are applicable to building types, other than residential dwellings and provides a refresher of what may be happening internally with the fire dynamics.

However, there are always exceptions to rules, so intelligence gathering, and pre-planning should be an essential part of fire station working routines.

"Big Box" Type Buildings.

These types of buildings are categorized together by their shared feature of a large "volume" or open space within them rather than their purpose group or use, though this should be considered when evaluating the likely fuels. Therefore, we may have a large retail showroom, a storage facility, manufacturing factory or even large open plan office buildings all categorized as "Big Boxes".

Air factors that are common to big boxes derive from the large geometric volume within the building. Air supply is plentiful and unlikely to be used up and put the fire into a "false decay" or situation where flaming combustion is not present. Given the large openings such as storefronts (retail), large roller shutter doors (factories) and possibly extensive glazing (open-plan offices), any failures or openings will ensure a supply of air to sustain combustion.

Fire profiles will generally be what we would simply categorise as "fuel controlled" (*ignoring the obvious question of whether any structural fire can be truly fuel controlled) and will be dictated by the fire loading available, which reverts back to knowing the building, and its contents. As we noted in Fire Dynamics for Firefighters, there is potential for fires throughout big boxes to be at different stages of development as the fire spreads due to the large volume, these have been attributed the description "travelling fires".

It is likely that there will a lower ratio of fuel to air in ballrooms, assembly halls and auditoriums, for example, and this may take fire gases out of their flammable range and become 'too lean to burn' in certain areas, while they remain within their flammable range in other areas.

Due to high ceiling/roof heights in "big boxes" hot fire gases can reach a certain height (below the ceiling) where they have started to cool, losing volume and buoyancy, then starting to sink back towards floor level.

"Big Boxes" present less opportunity for heat to transfer horizontally across ceilings (convection) and to radiate vertically back down into the compartment, so the speed of development may be less pronounced than small compartments due to lower "plume" heights.

However, there may be significant fire loading fuels available that can contribute to a massive heat release rate and a rapidly developing fire. Big boxes such as Supermarkets and warehouses with racking systems can cause a 'compartmentalising' effect, acting like walls and allowing horizontal radiant heat (at lower heights) to contribute to fire growth.

Artificial corridors/aisles created by racking can also create an effect like a corridor or pipe, creating a flowpath with less resistance, allowing fire and gases to travel along its route to the outlet/exhaust vent.

It is not out of the question that we can encounter smaller areas of 'localized flashover' within larger compartments.

In addition to the classic SAHF factors, Incident Commander should consider the following within their tactical decision making at Big Box incidents:

BA teams may be making entry at differing stages of fire development in contrast to a domestic dwelling fire. Firefighters may be entering in a developing fire stage, which has not flashed over. This can present more immediate danger than a fire which may have started to go into decay having used all available fuels.

In larger commercial buildings, heating and ventilation systems, ducting and pipework provide opportunities for hidden fire spread and travel, particularly if dampers and systems haven't been maintained correctly.

If cladded panels/ ACPs/ LISPs are used in the construction and fail in fire, there can be huge inlets, drawing in air, possibly wind driven, and all ingredients are present in a recipe for disaster. Cladded panels, potentially without cavity barriers can delaminate, conceal hidden and rapid fire spread whilst producing extremely thick and toxic smoke.

Larger volumes of compartment will require a larger volume of fire gas to fill before starting to become pressurised, so this may seldom, or never be seen.

External signs such as gases being pushed through gaps under pressure may not be exhibited.

Fire gases will also be present in different concentrations (flammability limits) in different areas in the compartment.

What is the fuel package and its likely effect on fire development?

Have artificial flow-paths been created by racking/aisles with potential for blowtorch effects?

- Multiple flowpaths can cause turbulent conditions and accelerate fire development.

High Rise Buildings (Commercial and Residential)

High Rise Firefighting could be a full book, or even series of volumes. As with any other building, pre-planning and knowledge of the building's layout and compartmentation, fixed installations (rising mains/standpipes, smoke extraction, pressurisation systems etc), evacuation strategies and exterior envelope should be knowledge that is immediately available and understood by responding fire crews.

Starting from the "outside in", with the building's "exterior envelope". The Incident Commander should consider whether the building is cladded with ACPs, LISPs or other types of cladding that can support rapid vertical fire spread. Prior knowledge is essential, such as whether the cladding systems (external wall systems- AKA EWS) have been correctly fitted to frames with "cavity barriers" which are design features to prevent vertical and horizontal fire spread. Smoke, Heat and Flame factors should be assessed, the use of a thermal image camera to detect heat signatures should be a primary consideration.

Commercial high-rise buildings may be more likely to have extensive glazing, which is invariably well insulated, when sizing up, this may impede the thermal image camera, also the height (distance to view), and time of day, may present difficulties in noting any heat signatures, blackening of windows, cracking or activation of lamination.

Naturally, a primary consideration at high rise fires, in addition to the external spread of fire, either supported by combustible cladding, or the "Coanda effect" is the potential for "Wind Driven Fires".

Whilst we explored this phenomena in "Fire Dynamics for Firefighters", we can recall that the explicit signature for a wind driven fire, that does not have a "completed" uni-directional flowpath is the "belching" appearance, where the pressure within a compartment is continually increased until it is higher than the exterior of the building, and temporarily overcomes the wind-driven effect, with a brief "belch" where the interior pressure is exhausted to the exterior- higher pressure gas moving to areas of lower pressure. Obviously, opening interior doors and allowing the wind to drive the fire through the building by "completing" a uni-directional flowpath creates a "kill-zone" or "fatal funnel".

Moving to the interior of high-rise buildings, where the Commander's SAHF assessment continues, fixed installations such as smoke extraction systems and pressurisation can affect the SAHF factors.

Fixed installation systems should, if maintained and installed to comply with "Building Codes" contain "fire dampers" which are situated where ducting or vent systems pass through compartment wall and shut to prevent fire spread, they may also be fitted with "smoke dampers" which open and extract or "draw" smoke in order to prevent accumulation at head height and effect upon persons escaping from the premises. Operation of these systems may lead to smoke being "drawn" some distance from its origin, and misleading SAHF factors exhibited.

A classic example of this can be where a pressurised enclosed stairwell also has a smoke extraction vent at the top of the shaft. The vertically rising airflow (flowpath) of fresh air inputted by the system that pressurises the stairwell can create a 'flue' with areas of negative pressure created as it passes each horizontal floor. Buoyant and pressurised fire gases will move to these areas of lower pressure with the potential for fire spread, and are sucked towards the stairwell, which may well be the point of our 'Bridgehead' or 'Forward Command Post'.

Compartmentation of high rise buildings should also be considered when evaluating SAHF factors. Rising and dropping service mains, broadband cabling may not have been correctly "fire-stopped" to the same level of fire resistance as the compartmentation walls and floor slabs. This may present potential for "leakage" of fire gases into adjacent compartments with FGI potential and can present false indicators of the origin of the fire. Of course, fire spread due to compromised compartmentation should be considered as part of the initial size-up.

Environmental factors, including socio-economic factors and people should be briefly taken into account as part of the Commander's size-up. These are particularly relevant at high rise incidents. Will (or has) the effect of persons escaping altered layouts and available "air" (doors left open or wedged open? A commonly encountered issue I experienced as a Commander in a Metropolitan area was Bicycles being chained over and under communal area doors, preventing these closing and compromising the fire protection of the stairwell or shaft. These can have an effect upon the fire dynamics within the interior of a high-rise building, which may not be apparent from the exterior.

In addition to the classic "SAHF" Factors, the Incident Commander should consider the following within their tactical decision making at High Rise incidents:

- **Exterior**- do any EWS present the potential for full-height rapid fire spread?
- **Exterior** – are any signs of "wind-driven" fire presenting?
- **Exterior**- can "Coanda Effect" impact upon this incident and fire spread?
- **Interior**- are fixed installations impacting upon fire development or presentation?

- **Interior**- will any "uni-directional flowpath" be completed and a "kill-zone/fatal funnel" created by opening any doors?
- **Interior**- have people factors altered or impacted upon internal layouts, with consequential effect on fire dynamics (more air supply due to open doors etc)?
- **Interior**- is compartmentation intact or is there potential for Fire Spread, Fire Gas Travel (FGI Risk) throughout the building?

B&B/Boarding Houses (Paying Guest Accommodation)

Hotels that are built to "code", are generally straightforward to fight fires in, they are well compartmented, generally well maintained and solidly constructed. However, there are other types of accommodation such as Boarding Houses, Bed and Breakfasts and Hostels that are generally converted from other uses. Many coastal B&Bs were formally residential properties that have been converted into use. Temporary accommodation that is increasingly offered upon the internet, of which *"Air BnB"* is the most well-known, may also be in domestic residences. These types of buildings may have been converted prior to building code requirements, or even without knowledge or approval of building control departments.

The arriving Commander can experience certain difficulties at these type of buildings. To maximise income, additional partitions may have been installed within these type of buildings, creating additional bedrooms. If installed correctly, then these can result in a limited source of air-supply to a fire, with the potential for "smoulder" or non-flaming combustion, release of pyrolysis products/unburned fuels. This lends itself to more of a risk of backdraft or fire gas ignition as opposed to flashover. However, if partitions and compartmentation are not installed correctly, there could be concealed fire spread within the building. Human factors such as persons leaving the building, can also alter the fire dynamics by creating more openings and potential air supply.

The presence of half-landings, additional and unexpected partitions and convoluted corridors and internal layouts can create voids, with the risk of "leakage" and fire gas ignitions. This also results in "flowpaths" being complex, potentially creating turbulence, affecting development.

Upon arrival, the Commander's 360 degree assessment is absolutely vital, as is the use of the thermal image camera as part of this. It can be difficult to evaluate the room of origin, if the building has a certain depth, or is part of a "terrace" or a "rowhouse". A number of fire escapes that persons have exited from, if left open, can provide air supply to the fire, or complete flowpaths, windows may have been opened by guests/residents and SAHF factors may be exhibited some distance from the origin of the fire.

Naturally, the classic SAHF factors for backdraft such as blackened windows, smoke under pressure or pulsating from gaps, remote fire gas ignition (diluting down outside the compartment and igniting – suggesting a too rich concentration of fuel, above its auto-ignition temperature within) should all be considered as possible factors at these type of buildings.

In addition to the classic "SAHF" factors, the Incident Commander should consider the following within their tactical decision making at Guest Accommodation building incidents:

Exterior- have multiple fire escapes/windows been opened creating inlet flowpaths and potential turbulence?

Exterior- are multiple heat signatures present when scanning with a thermal image camera?

Exterior- smoke colouration and pressure may indicate unburned fuels suggesting limited ventilation profile and backdraft/FGI potential. Remote ignitions would also support this

Interior- have people factors altered or impacted upon internal layouts, with consequential effect on fire dynamics (more air supply due to open doors etc)?

Interior- is compartmentation intact or is there potential for Fire Spread, Fire Gas Travel (FGI Risk) throughout the building?

Heritage and Traditional Buildings

Existing heritage buildings are usually of historic importance. They may be larger, such as mills and share various features with "big box" type buildings. Pre-20th century buildings, and also certain 20th century buildings, can affect fire dynamics. If we look at the materials in heritage properties such as old libraries, museums, castles, mills stately homes, the contents may be dryer and less humid in order to preserve them.

With less moisture in the fuel, less water vapour and carbon dioxide are released to act as passive agents and absorb energy to slow fire development. As you may expect, this can lead to faster developing fires with fuel conditions that contribute to effective combustion.

However, the opposite can also be true. Heritage properties can also be damp and therefore have more water based pyrolysis products acting as passive agents and slowing the fire development.

The Commander can utilise the construction of the building to effectively aid the "size-up". Heritage and traditional buildings will often have an exterior consisting of brick or stone which will absorb heat and be picked up on a thermal image camera. Where glass is present, it may be in single pane format with wooden frames which also will exhibit heat. Heritage buildings therefore may have a number of potential "flowpath" inlets as part of the "air factors" if glazing fails.

If the buildings pre-date "codes" then there may be voids and compartmentation may be limited- exposed joists and flooring boards, especially if affected over years by rot, termites or woodworm, this can cause fire spread internally and sources of "leakage" with both fire spread and FGI risks.

Compartmentation in old buildings can consist of "lathe" and horse-hair (to provide insulation)- should a fire breach these voids, horse-hair will sustain a slow, smouldering burn, if air supply is limited. This may produce strange signatures, with localised FGI potential within voids.

Buildings such as former bonded warehouses, distilleries and manufacturing plants present specific risks. Materials such as alcohols, solvents and chemicals may have impregnated or been absorbed into the linings such as walls and floors over a number of years, and in a fire situation, release numerous chemical combinations forming gaseous fuels. This can seriously affect the speed of fire development particularly in gaseous phase combustion.

Heritage properties have a number of potential effects on fire behaviour and traditional construction increases potential for collapses. Cast iron columns, commonly used in heritage buildings, can be subject to "thermal shock" and shatter, causing floor collapse, if struck by a cold water jet. Any partial collapse can create large openings and provide more air to the fire, increasing the rate of development.

Windows have all got the potential to become inlets, affecting the ventilation profile of the fire and increasing the potential for turbulent gas flows. This makes interior attack/compartment firefighting extremely challenging and hazardous.

Layouts, fire loading /fuels within heritage buildings should all be considered as part of the pre-planning phase.

In addition to the classic "SAHF" factors, the Incident Commander should consider the following within their tactical decision making at Heritage building incidents:

Exterior- do any openings or windows create inlet flowpaths and contribute to potential turbulence and faster fire development?

Exterior- are heat signatures visible in masonry, brickwork or stone when scanning with a thermal image camera?

Exterior- smoke colouration and pressure may indicate unburned fuels suggesting limited ventilation profile and backdraft/FGI potential. Remote ignitions would also support this

Interior- does the fire loading/ available fuel or internal layouts have any consequential effect on fire dynamics (more air supply due to open doors etc)?

Interior- is compartmentation intact or is there potential for Fire Spread, Fire Gas Travel (FGI Risk) throughout the building, particularly voids and walls?

Modern Buildings and Fire Engineering Impacts

Whilst we have touched upon the impact of smoke extraction, pressurisation and their potential consequences upon fire development, and the subsequent presentation of signals and impact on the Commander's decision making, we should also assess modern buildings in their "passive context".

As the planet moves towards carbon neutrality, buildings are beginning to become better insulated in order to reduce reliance upon heating systems within. This means that construction products with better insulation and integrity are more frequently used in buildings. Some European buildings are now "leakage tested" to ensure that they do not lose heat, in the form of warmed air, from the building, and retain heat more efficiently. This efficiency obviously has impact for the Commander's assessment.

Whilst assessing the SAHF factors, the building is designed to keep the heat within. If we consider our heat transfer methods, the walls, glazing, roof structure of the eco-efficient buildings are designed not to conduct heat, so we may not see any exterior heat signature, that we may do on traditional construction. Consider the "leakage" test, the building design reduces or removes any gaps, so the convection cycle is affected- the Commander may not see any pressured gases being forced from gaps, as these have been eliminated from the design. Naturally, the uncontrolled movement of air into the building is reduced, which impacts upon the fire dynamics. (Air will still be drawn into and exhausted the building, but this shall be by design).

This methodology is not just used to keep buildings warm, but in hotter climates, to also increase efficiency in the maintenance of coolness, which reduces air-conditioning costs and environmental impact.

Modern, efficient buildings can prevent effective size up using the SAHF factors, due to the reasons detailed above; smoke and air may not leak (both into and out of the building), heat signatures may not be conducted and may be less visible to a thermal image camera.

Fire development will depend upon layouts, fuel loading etc, but the high degree of insulation and lack of uncontrolled airflow will lend itself towards backdraft and a fire becoming ventilation controlled earlier in its stages of development.

In addition to the classic "SAHF" factors, the Incident Commander should consider the following within their tactical decision making at building incidents with modern fire engineering solutions

Exterior- do any emissions, such as smoke from open vents, or remote ducting from fires indicate that any smoke control systems are in use?

Exterior- is this an eco-efficient building with insulation that impedes thermal image use or the assessment of the classic SAHF factors?

Interior- are smoke control systems or pressurisation having an adverse effect upon Fire conditions such as creating lower pressure areas for fire to spread into?

Interior- can smoke control, pressurisation and extraction systems be

Summary: Building Types:

Effective SAHF assessments always factor in that the construction features of a building may impact upon what can be seen from the exterior of the building. Whilst we know what we "should" be seeing and what these indicators, when present, tell us, the effective commander does not discount the effect that particular building types can impact and provide "misleading information".

At risk of "labouring the point", the importance of knowing the buildings and construction that are present in one's geographic fire station area has never been so important. Most facilities management companies and construction firms are willing to engage in positive dialogue with operational responders and provide plans and tours of buildings and it's features, so be proactive and don't wait for an emergency to be the first visit to a building.

Chapter 12: Reading the fire – a summary

Well, we have made it to the end of this volume. Before we sign off though, and move onto Volume 3: Fighting Fire, let us have a quick reminder of how the BE-SAHF indicators give us valuable clues to the stage of a fire's development, and its likely progression.

Firefighters are expected to work in dangerous and rapidly changing environments. They must make decisions in seconds, with very limited information, and take actions that may save, or endanger, lives and property. The best we can do is to base our decisions on a sound knowledge of building construction, and the ability to recognise the fire behaviour patterns at each stage of the incident. There are no easy answers, but as scientific research enlightens our knowledge of fire behaviour, we must be prepared to challenge our traditional thinking and be prepared to open our minds to new possibilities.

Reading the fire is like learning a language. Every fire is 'speaking' to us through the fire behaviour indicators. Sometimes the fire is talking softly, and even if we understand the language, we must be paying close attention, or we will miss the message. Sometimes the indicators are so clear that the fire is shouting at us, but if we don't understand the language, then it doesn't matter how loud, or clear the message is. We must be aware that sometimes the fire tells lies. It deceives us by only telling part of the story, by concealing critical information.

By identifying several fire behaviour indicators, we will be able to recognise patterns. These may provide "soft", or "hard" information. As we continue to recognise the emerging patterns, we are gradually able to improve our situational awareness.

External observations of the building can never be relied upon to fully prepare us for what will be encountered, Therefore, not only must we know the language of fire, but we must also be aware that it will never give us all the "pieces of the puzzle".

But what is the alternative? Give up and rely on routine and guesswork? We are very confident that the ability to recognise the emerging fire behaviour patterns can do much better that this "pedestrian" approach!

Never rely on one indicator! Read them all in the context of the building factors and ambient environmental factors. All firefighters should be observing the FBI in their area of operation and should relay this critical information to the incident commander. This is process is dynamic and must continue until the fire ground is vacated.

In the next book, we will cover how we can utilise our knowledge of fire dynamics, and our ability recognise the fire behaviour indicators, as a foundation to develop the most effective and safe strategies and tactics.

We can move beyond "routine and guesswork" We must stop relying on the 1 Tactic, 1 Tool and 1 Technique approach.

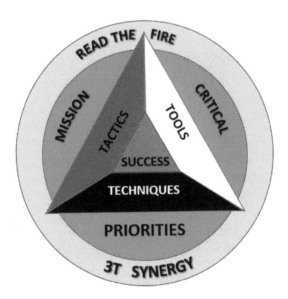